AF540431

# A TEXT BOOK OF

# FUNGI

# A TEXT BOOK OF
# FUNGI

**Dr. Rajni Gupta**

**APH PUBLISHING CORPORATION**
7, ANSARI ROAD, DARIYA GANJ, NEW DELHI-110 002

*Published by*
S.B. Nangia
**A P H Publishing Corporation**
7, Ansari Road, Daryaganj
New Delhi 110002
Ph.: 23274050
E-mail : aphbooks@gmail.com

2026

Rs. 3995/-



*Printed at*
**Balaji Offset**
Navin Shahdara, Delhi 110032

# Preface

This book of Fungi has been planned for the undergraduate students but an effort to make it useful to honours and post graduate students has also been made. There are some good books available on the subject-both by foreign and Indian authors but they are either too elobrate and beyond the scope of an average student or they are too elementary and lack in important details to give a fair idea of the subject to the students. Consequently, the author felt the need of yet author book which could present before them the broad range of variation in structure and life history in a simple manner and present work is an effort in this direction.

There is a advancement of knowedge in all branches of science including fungi and the subject could be approached from various angles i.e. physiological, ecological, morphological, pathological and on molecular genetics basis. This book is based on morphology of fungi and some pathological aspects dealing with important crop diseases. The classification of fungi, like any other groups is quite controversial at various stages, but it has been given in a simple way following the latest based on DNA sequence analysis.

Syllabus of Delhi University and other Universities have been consulted and effort has been made to include as many representatives types as possible to increase the utility of this book. Empahsis has been given on fungal biotechology also. Different types of microbes are extensively used in the production of a large variety of proteins, enzymes vitamins, antibiotic, steroids, pharmaceuticals, food ingredients, alcohols and biocides.

Mycorrhizal fungi are importnat tools for increasing growth, development and yield of economically important plants. They play important role of biofertilizer which can help establish plants in nutrient deficient soil. They help in afforestation also.

It would not have been possible to complete this work without the help of many friends and well wishers. I am highly thankful to Prof. K.G. Mukherji, Dept. of Botany, University of Delhi, Delhi, who helped me in various ways. I am also thankful to Dr. B.P. Chamola for provding help in various ways during the preparation of this book.

This book will be useful to both undergraduate and post graduate students.

**Rajni Gupta**

# Contents

# 1. INTRODUCTION

Fungus is Latin word which means "**mushrooms**". In usage meaning of the word, has been expanded to include thallus like, non green plants such as **molds, yeasts** and other similar **organisms** related to mushrooms. Thus fungi are a large group of simple, thallus like plants which lack chlorophyll. The earlier botanists employed the term '**fungi**' in a wider sense to include **bacteria, slime molds** as well as true fungi (**Eumycetes**).

The branch of botany which deals with fungi is called **mycology**. In Greek **Mykes** means mushroom and **logos** means science which deals with the **study of mushrooms**. It deals with life histories, relationships and evolutionary tendencies of fungi. The role of the edible fruit bodies is the production of large number of spores by means of which dispersal occurs. Fruiting body is composed of long, cylindrical, branched thread like structures *i.e.* **hyphae** (*Sing., hypha*). The hyphae are divided by cross walls into compartments which typically contain several nuclei. Such compartments, together with their walls, are equivalent to the cells of other organisms. The spores are borne on specialized cells termed **basidia** (*Sing, basidium*). In most mushrooms each basidium bear's four spores, but in ***Agaricus bisporus*** only two, hence the specific epithet *bisporus*.

**Fungi: A Major Group of Organisms**

Different kinds of large fungi or **macrofungi** have been recognized for thousands of years. In current English language edible ones are often called mushrooms and **poisonous** ones **toad** stools. During the eighteenth century, botanists made considerable progress in the recognition and classification of the fungi, and early microscopists observed and described hyphae and **spores**. In the early nineteenth century, it was established that many serious plant diseases were caused by **infection** of the plant by minute living organisms. These **organisms** were found to be composed of hyphal branches which were capable of producing spores. Several plant pathogens, such as **rusts, smuts** and **mildews** were recognized as microfungi. Microfungi were also found growing on dead organic materials. Such microfungi were termed as **saprophytic moulds** (molds).

Studies on **alcoholic fermentation** established that the **yeast** responsible for the process was a microscopic organism reproducing by **budding**. Although the cells of Brewer's yeast are ellipsoidal in form and do not produce hyphae, they were regarded as fungi on the basis of being plants that live by absorbing organic materials rather than by **photosynthesis**. The discovery of yeasts capable of producing a hyphal phase and some moulds able to produce a yeast phase confirmed the soundness of classifying yeasts as **fungi**, even though some of the best known species did not form hyphae. Thus fungi can hence be defined as non-photosynthetic hyphal **eukaryotes**. It includes eukaryotic, spore bearing, achlorophyllous organisms that generally reproduce sexually and asexually, and those usually filamentous, branched somatic structures are typically surrounded by cell walls containing chitin or cellulose, or both of these substances, together with many other complex organic molecules.

## General Characters

The fungi constitute a group of living organisms devoid of chlorophyll. They resemble simple plants in that, with few exceptions, they have definite cell walls, they are usually **non motile**, although they may have **motile** reproductive cells, and they reproduce by means of spores. A **spore** is a minute, simple propagating unit without an embryo that serves in the production of new individuals of the same species. Fungi do not possess stems, roots or leaves, nor they have developed a vascular system, as plants have. Fungi are usually filamentous and multicellular, their nuclei can be demonstrated with relative ease, their somatic structures, with few exceptions, exhibit little differentiation and practically no division of labour. The filaments constituting the body of a fungus elongate by apical growth , but most parts of an **organism** are potentially capable of growth, and even minute **fragments** from any part can give rise to a new individual. Reproductive structures are differentiated from somatic structures and exhibit a variety of forms and shapes.

## Habitat

The fungi grow in diverse habitats. In fact they are found in almost every available habitat on earth where organic material (living or dead) is present. They are thus universal in their distribution. Generally many of them are **terrestrial**. They occur in soil which abounds in dead, decaying organic materials. The terrestrial fungi thrive best in humus soil. They are considered more advanced. They produce **non motile** reproductive cells which are dispersed passively by wind, water or animals. Some fungi attacks living organisms. They live in tissues of plants and animals. Some fungi are **aquatic**. The aquatic fungi are considered primitive. They live on decaying organic matter and living organisms found in fresh water and produce flagellate reproductive cells which swim to new localities. Many fungi grow on our food stuffs such as bread, jams, pickles, fruits and vegetables. Some fungi are found in drinking water. They are present all the time in air that we breathe. Majority of these prefer to grow in darkness and dim light in moist habitats.

## Nutrition

In nature fungi obtain their food either by infecting living organisms as **parasites** (Gr. *Parasitos* = eating beside another) or by attacking dead organic matter as **saprobes** (Gr. *Sapros* = rotten + *bios* = life). Many also form symbiotic relationships with plants as in lichens and mycorrhizae. The majority of known fungi, whether parasitic or not, are capable of living on dead organic material, as shown by their ability to grow artificially on synthetic media. Fungi that live on dead matter and are incapable of infecting living organisms are called **obligate saprobes**; those that are capable of causing disease and can also live on dead organic matter are known as **facultative parasities**, and those that can not live except on living protoplasmas are referred to as **obligate parasites**. A living organism infected by a parasite is known as the host. Fungi capable of growing parasitically but can also survive saprobically on dead organic matter are according to circumstances, referred to as either **faculative parasites** or **facultative saprobes**. In some cases fungi live either with in or on host animals without causing any obvious damage to the animals. In other fungi actually trap and consume microscopic animals as food. Some of these so-called **predacious** species produce elaborate structures that function in the capture of their prey. Fungi require elaborate food in order to survive since they are achlorophyllous and are incapable of manufacturing their own. They prefer carbohydrates in form of glucose or maltose. Most fungi can synthesize their own proteins by utilizing inorganic or organic sources of nitrogen and various mineral elements essential for their growth. Many fungi are capable of synthesizing vitamins. Some, however are deficient in thiamine or biotin or both, and must obtain these or their precursors from the substratum. Fungi usually store excess food in the form of glycogen or lipids.

Different fungi have different food requirements. Some are omnivorous and can subsist on anything that contains organic matter. The common fungi ***Aspergillus*** sp. given a little moisture, will grow on anything in their diet; a few obligate parasites not only require living protoplasm for food, but also are highly specialized as to the species and even the variety of host they parasitize.

Fungi generally prefer food in soluble state, the food molecules must be of small, enough size to be capable of passing freely through the cells walls and membranes. Thus a fungus must break down larger molecules into smaller ones before it can absorb them. It does this by secreting extracellular enzymes that act on the substratum digesting the food outside the fungal body. Thus fungi can produce large amounts of enzymes and so they can use different type of substrates as food.

**Somatic Structure**

The fungal thallus typically consists of microscopic thread like structures filaments that branch in all directions, spreading over or with in the substratum utilized as food. Each of these filaments is known as **hypha**. A hypha is made of a thin transparent, tubular wall filled or lined with a layer of protoplasm varying in thickness. (Fig. 1.1A)

The protoplasm in the hyphae of most filamentous fungi is interrupted at irregular intervals by **partitions** or cross walls that divide each hypha into compartments or cells. The cross walls are called **septa** (Sing, septum) (Fig. 1.1B). In the simpler filamentous fungi septa are always formed at the base of reproductive organs but are confined to those locations so that the vigorously growing hyphae are coenocytic and non septate (aseptate). Even here, however, when the hyphae septa are often formed at various places. As portions of a hyphae die and the protoplasm is withdrawn the growing tip, a septum that separates the dead portion from the living is generally formed.

In septate forms, depending upon the species, each all may have either one nucleus or two or more nuclei. Accordingly, the hyphae would be designated as uninucleate, binucleate and multinucleate. When the hypha is not septate and it contains numerous nuclei it is called **coenocytic**.

In some primitive forms, the thallus may be represented by a single cell or it may contain a small rhizoidal system or haustoria. In case, a plant body consists of a single cell and is converted completely into a reproductive structure is known as **holocarpic**, but if only a part of the thallus is used up in the formation of reproductive structure, it is called as **eucarpic**.

Generally, the mycelium consists of a net work of loose hyphae aggregated together and growing on or within the substratum. In most of the higher forms during certain stages of the life history, the mycelium gets organised into a loosely or compactly woven tissue like structure. These are termed **Plectenchyma**. Two types of Plectenchyma are:—

(i) **Prosenchyma:** In which the hyphae are loosely inter woven and lie more or less parallel to each other so that longitudinally oriented hyphae and cells are clearly distinguished (Fig. 1.1C).

(ii) **Pseudoparenchyma:** In which the hyphae are closely packed and interwoven so that completely loose their identity, appear as an isodiametric cells giving the tissue an appearance of parenchyma found in higher plant groups (Fig 1.1D).

We recognize two general types of septa namely **primary** and **adventitious**. The **primary septa** are formed in association with **nuclear division** and are laid down between daughter nuclei. The **adventitious septa** are formed independently of **nuclear division** and are especial association with changes in the concentration of the protoplasm as it moves from one part of the hypha to another (Talbot, 1971). Septa vary in their construction, some are simple, other more complex. All types appear to be formed by centripetal growth from the hyphal wall inward. In some septa growth continues until the septum is a solid plate; in others the septum remains incomplete; leaving a pore in the center that may often by plugged or occluded. In the most complex fungi the septa have a special central apparatus in the form of barrel shaped inflation surrounded typically by a perforated membrane. This is known as **dolipore septum** (Moore and Mc Alear, 1962).

**Plasmodesmata** have also been demonstrated in a few fungi. They are probably of more common occurrence than observations to date indicate. The individual cells of septate hyphae may contain one, two or many nuclei. **Uninucleate** cells are characteristic of some fungi, **binucleate** cells of others, whereas, **multinucleate** cells may occur in most.

## Cell Wall Composition

The cell wall is multilaminate, with the **lamellae** consisting of variously oriented fibrils (Aronson, 1965). The principal chemical constituents of the fungal cell wall are various **polysaccharides**, but **proteins, lipids** and other substances are also included. The

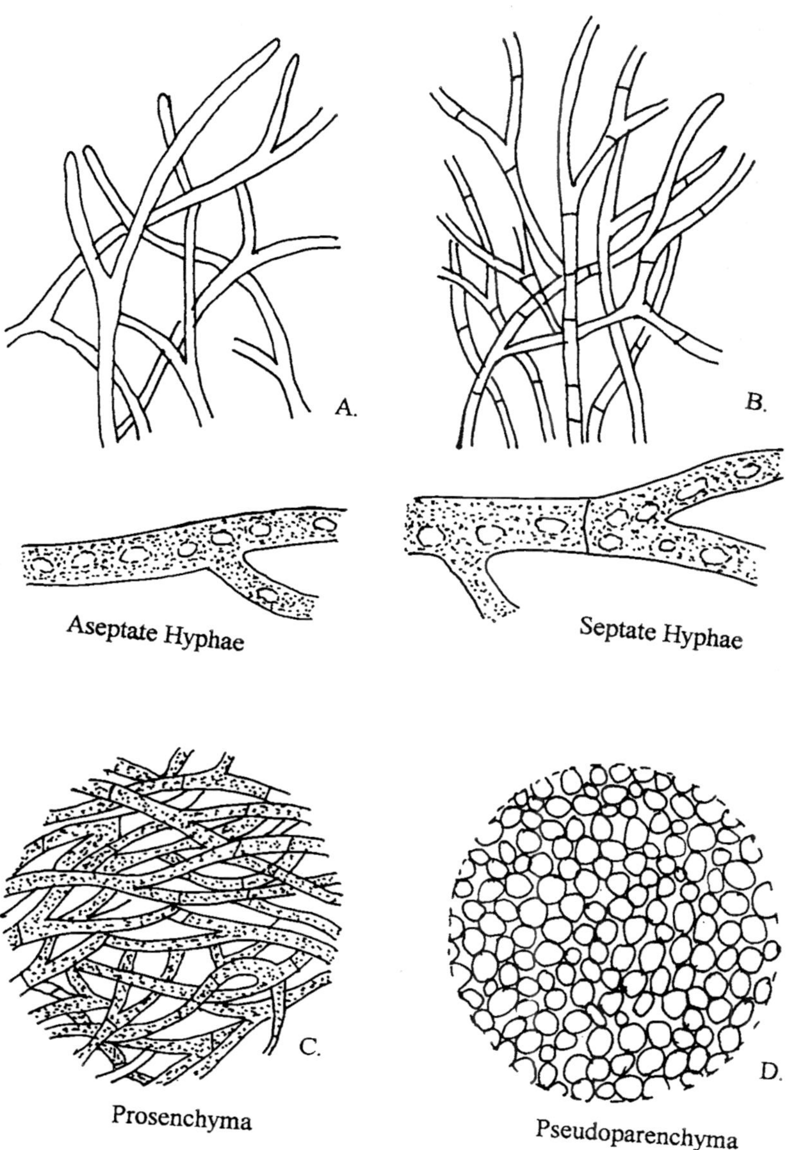


**Fig. 1.1:** Fungal hyphae and fungal tissues

chemical composition of the cell wall is not the same in all fungi, and also varies with the change in circumstances. On the contrary, substance that may be present in the young hyphae may disappear almost completely as the hyphae grow older, or other materials may be deposited and mask the presence of the earlier constituents, making their detection very difficult. Further more, it has been shown definitely that external factors such as the composition of the media, pH values, and temperature profoundly influence the composition of the fungal walls (Foster, 1949).

**Mycelium**

The mass of hyphae constituting the thallus of a fungus is called the mycelium (pl. mycelia). The mycelium of some fungi forms thick strands. In certain types of such strands – the **rhizomorphs** (Gr. *Rhiza* = root + *morphe* = shape) – the unit hyphae lose their individuality and form complex tissues that exhibit a division of labour. The string like mass has a thick, hard cortex, and a growing tip whose structure reminds us of that of root tip. Rhizomorphs are resistant to adverse conditions and remain dormant until favourable conditions return (Fig. 1.2 A–C). Growth is then resumed, and the rhizomorph attains great length. **Rhizomorphs** are usually produced by the most complex Basidiomycetes, but are also found in other groups (Goos, 1962). Hyphae of plant pathogenic fungi growing with in the tissues of their hosts exhibit various patterns of growth depending upon the type of pathogen involved (Nims, 1991). Basically plant pathogenic fungi fall into three categories. Perthrotrophs also known as **necrotrophs**, use enzymes and toxins to kill host cells in advance of their hyphae and then grow between and into dead and dying cells. **Biotrophs** are ecologically obligate parasites and invivo obtain nutrients only from living host cells. Third category of plant pathogenic species are the **hemibiotrophs**. Hemibiotrophs initially require living host cells but soon cause the death of host cells in advance of their hyphae like the perthotrophs noted above. The mycelium of parasitic fungi grow on the surface of the host either spreading between the cells or penetrating into them. If the mycelium is intercellular, food is absorbed through the host cell wall or membrane. If the mycelium penetrates into the cells, the hyphal walls come into direct contact with the host protoplasm. Intercellular hyphae of many fungi, especially or obligate parasites of plants, obtain nourishment through **haustoria** (Sing, haustorium). Haustoria, through which the fungus sinks into the plant host cells by a minute pore punctured in the cell wall, are out growths of the somatic hyphae. They are regarded as specialized absorbing organs (Bhandari and Mukerji, 1993). **Haustoria** may be knob like in shape, elongated, or branched like a miniature root system. The production of haustoria is probably in response to the contract stimulus as well as to the stimulus of nutrients. Haustorium like branches of the mycelium are also produced by the black stem rust fungus when growing axenically on agar under conditions unfavourable for saprobic growth (Williams, 1969). The hyphae of saprobic fungi come in intimate contact with the substratum and obtain food by direct diffusion through the hyphal walls, causing disintegration of the organic matter that they utilize. The older hyphae die as the mycelium grows and branches; and they themselves disintegrate in nature because of the activities of other microorganisms that feed on their dead bodies.

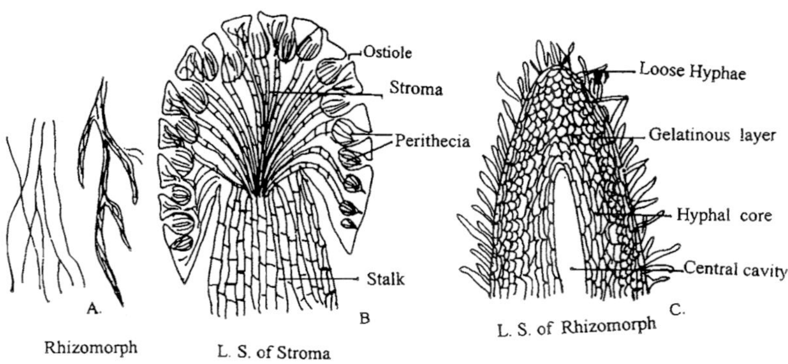


**Fig. 1.2:** Rhizomorph and stroma

There are three major growth forms of fungi. Most fungi are mycelial, with a network of hyphae, and colloquially they are termed '**moulds**'. Some of the more primitive fungi, such as the chytridiomy cota often have single rounded cells or dichotomously branched chain of cells, attached to a food source by tapering rhizoids. Some other fungi grow as unicellular yeasts and produce daughter cells either by budding (*Saccharomyces cerevisiae*) or by binary fission. These different growth forms can be related to habitat. For example, yeasts are common in most environments that are rich in simple, soluble nutrients, and some fungi can even alternate between a yeast and a mycelial phase response to environmental conditions. Such dimorphic fungi include some of the important pathogens of humans and higher animals they grow as yeasts spread in water films or in body fluids, but as hyphae for invasion of the tissues *Candida albicans* is a classic example of this. It is common as a yeast on the mucosal membranes of humans, where it does little damage; but in response to some conditions the yeast cells can produce hyphae which invades the mucosa and cause significant problems.

Hyphae of plant pathogenic fungi growing with in the tissues of their hosts exhibit various pattern of growth depending upon the type of pathogen involved. Basically plant pathogenic fungi fall into one of three categories. **Perthrotrophs** also known as necrotrophs, use enzymes and toxins to kill host calls in advance of their hyphae and they grow between and into dead and dying cells. **Biotrophs** are ecologically obligate parasites and obtain nutrients only from living host cells. In case of Biotroph, hyphae primarily grow between host cells and give rise to special hyphae branches that penetrate the host cell without killing them **Hemibiotrophs**, they initially requires living host cells but soon cause the death of host cells.

**Growth**

Most fungi grown between 0° to 35°C, but the optimum temperature for their growth is 20° to 30°C. There are a number of thermophilic fungi which grow best at or above 50°C and at a minimum or 20°C or above (Cooney and Emerson, 1964). The ability of fungi to withstand extremely low temperatures, in a dormant state is utilized in the long term storing of fungal cultures in liquid nitrogen at a temperature of 19°C.

Fungi prefer an acidic medium for growth, with a pH of 6.0 being near the optimum for most species investigated. Although light is not required for the growth of fungi, some light is essential for sporulation in many species (Cochrane, 1958). However, it is interesting that the effect of light on some species appears to be socialized and is not transferred through the mycelium to nonilluminated portions of the thallus (Koehn, 1971). Fungal hyphae are capable of indefinite growth under favourable conditions. In nature, fungal colonies have been known to continue growing for 400 years or more. The mycelium of a fungus generally begins as a short germ tube emerging from a germinating spore. The mycelium has a tendency to grow more or less equally in all directions from the central point, and to develop a round colony.

**Fungal Metabolites**

The metabolites of fungi can be grouped into two broad categories (*i*) **primary metabolites** which are the end products of the common metabolic pathways and (*ii*) **secondary metabolites** which are a diverse range of compounds formed by specific pathways of particular organisms. Both types of product accumulate in the culture medium when growth is restricted by some factor but when the biochemical machinery continues to operate, like the engine of a car taken out of gear (Gupta & Mukerji, 2000). This is done purposefully in commercial conditions. For example, ethanol accumulates as a metabolic end product when yeast is grown in a sugar rich medium favouring metabolism, but in anaerobic conditions that limit growth, and some fungi produce large amounts of organic acids from sugar when their growth is limited by low pH. Many other primary metabolites can be produced by fungal fermentations, e.g. glycerol, mannitol, isomeric acid and $\alpha$-ketoglutaric acid, although the use of fungi to produce them commercially may not be economically feasible if they can be produced more cheaply by chemical synthesis.

A vast range of secondary metabolites are produced by fungi and are described by Turner (1971). Many of them have no commercial importance but other are extremely important (Table 1).

**Table 1:** Fungal Secondary metabolites produced commercially for pharmaceutical, agricultural and research uses.

(*Source* Deacon, 1997).

| *Usage* | *Product* | *Fungal Source* | *Application* |
|---|---|---|---|
| Medicine | Penicillins | *Penicillium chrysogenum* | Antibacterial |
| | Cephalosporins | *Cephalosporium acremonium* | Antibacterial |
| | Griseofulvin | *P. griseofulvum* | Antifungal |
| | Fusidin | *Fusidium coccineum* | Antibacterial |
| | Cyclosporin | *Trichoderma polysporum* | Immuno suppressant |
| | Ergot alkaloids | *Claviceps purpurea* | Induces labour: migraine treatment. |
| Agriculture Research | Zearalenone | *Gibberella zeae* | Growth promoter for Cattle plant hormones |
| | Gliotoxin | *Trichoderma virens* | Immunosuppressant |
| | Ctochalasins | *Helminthosporium dermatoideum* | Anti actin agents |
| | Fusicoccin | *Fusicoccum amygdali* | Stomatal opening |
| | Phalloidin | *Amanita phalloides* | Actin binding |
| | α-Amantin | *A. phalloides* | RNA polymerase binding inhibitor |

## Reproduction

Reproduction is the formation of new individuals having all the characteristics typically of the species. Two general types of reproduction are recognized: sexual and asexual. **Asexual** reproduction, sometimes called somatic or vegetative does not involve the union of nuclei, sex cells or sex organs. Sexual reproduction, on the other hand, is characterized by the union of two nuclei. In the formation of reproductive organs, either sexual or asexual, the entire thallus may be converted into one or more reproductive structures, so that somatic and reproductive phase do not occur together in the same individual. Fungi that follow this pattern are called **holocarpic** (Gr. *holos* = whole + *karpos* = fruit).

In the majority of fungi, however, the reproductive organs arise from only a portion of the thallus, while the remainder continues its normal somatic activities. The fungi in this category are called **eucarpic** (Gr. *eu* = good + *karpos* = fruit). The holocarpic forms are, therefore less differentiated than the eucarpic. Hennebert and Weresub (1977) proposed a naming system that has become widely accepted. In this system the term **telomorph** is used to describe the sexual stage of a fungus, while the term **anamorph** is used for its asexual stage. The holomorph is used to describe the whole fungus in all its forms.

## Asexual Reproduction

Typically, fungi reproduce both asexually and sexually. In general, asexual reproduction, is more important for the propagation of the species because it results in the production of numerous individuals, and particularly since the asexual cycle is usually repeated several times during the season, whereas the sexual stage of many fungi is produced only once in its life time.

Some fungi employ fragmentation of hyphae as a normal means of propagation (Fig. 1.3E). The hyphae may break up into the component cells that behave as spores. The spores are known as **arthrospores** (Gr. *arthron* = joint + *spore* = seed). If the cells become enveloped in a thick wall before they separated from each other or from other hyphal cells adjoining them, they are often called **chlamydospores** (Gr. *Chlamys* = mantle + *spora* = seed, spore). Fragmentation may also occur accidentally by the tearing off the parts of mycelium which under favourable conditions develop into new individuals. (Fig. 1.3D).

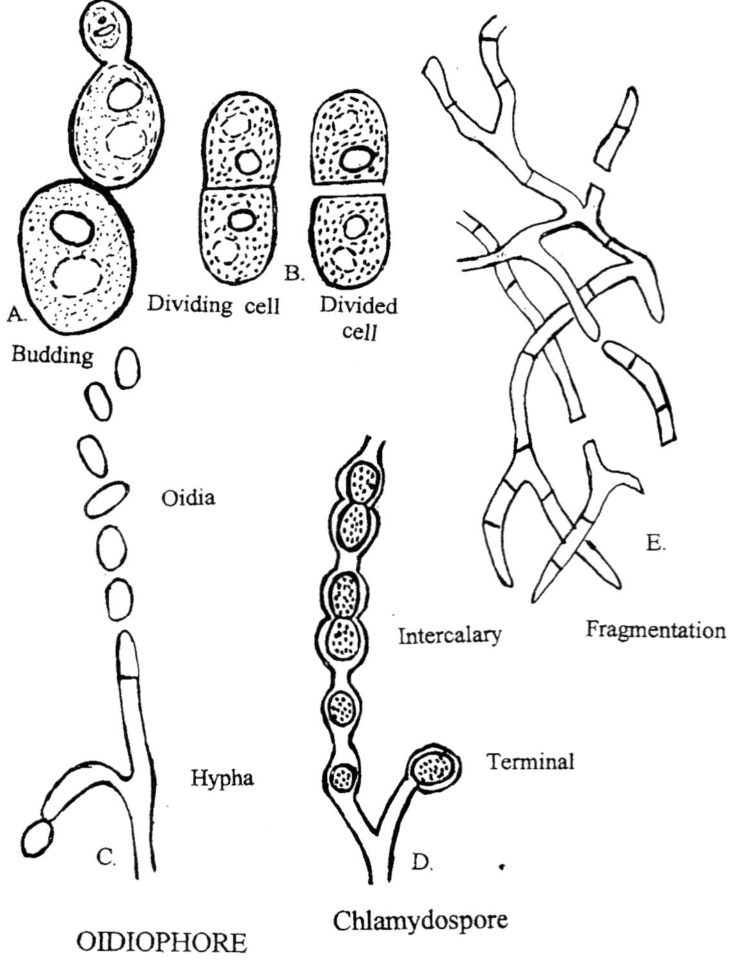


**Fig. 1.3:** Asexual Reproduction

Budding is the production of small out growth bud from a parent cell (A). As the bud is formed, the nucleus of the parent cell divides and one daughter nucleus migrate into the bud. The bud increases in size while still attached to the parent cell and eventually breaks off and forms a new individual. Chains of buds, forming a short mycelium, are sometimes produced. Most common mode of asexual reproduction in fungi is by means of spores. Spores vary in color from hyaline, through green-yellow, organge-red, brown to black; in size from minute to large and of different shapes, from globose through oval oblong, needle shaped to helical; in number of cells, from one to many; in the arrangement of cells; and in the way in which the spores themselves are borne. **Asexual** spores may either be borne in **sporangia** (Sporangium; Gr. *Spora* = seed, spore and *angeion* = vessel) and are then called **sporangiospores**, or are produced at the tips or sides of hyphae in various ways and are then called **conidia** (Sing, conidium Gr. *Konis* = dust and *idion* = dimim). Sporangium is a sac like structure whose entire contents are converted through cleavage into one or more-usually many, spores. Sporangiospores may be motile or non motile. In the simpler fungi the sporangiospores are usually motile and are called **zoospores** (G. *Zoon* = animal and *spore* = seed, spore). Zoospores are provided with one or two flagella. There are at least two types of flagella in the fungi: the whiplash and the tinsel shaped.

**Sexual Reproduction:** Sexual reproduction in fungi as in other living organisms involves the union of two compatible nuclei. The process of sexual reproduction typically consists of three distinct phases. In the first of these, called **Plasmogamy** (Gr. *plasma* = a molded object, *i.e.* a being + *gamos* = marriage, union), a union of two protoplasts brings the nuclei close together with in the same cell. The fusion of two nuclei brought together by plasmogamy is called **karyogamy** (Gr. *karyon* = nut, nucleus + *gamos* = marriage) and constitutes the second phase of sexual reproduction. Karyogamy follows plasmogamy almost immediately in many of the simpler fungi. While, in the case of complex fungi, however these two processes are separated in time and space, with plasmogamy resulting in a binucleate cell containing one nucleus from each parent. Such a pair of nuclei is known as dikaryon. In the later stage these two nuclei fuse, mean while, during growth and cell division of the binucelate cell, the dikaryotic condition may be perpetuated from cell to cell by the simultaneous division of the two closely associated nuclei, and by the separation of the resulting sister nuclei, into the two daughter cells. Nuclear fusion, which eventually takes place in all sexually reproducing fungi, is sooner or later followed by meiosis, which again reduces the number of chromosomes to the haploid and which constitutes the third phase of sexual reproduction.

Sex organs of fungi are called gametangia. These may form differentiated sex called gametes or may contain instead one or more gamete nuclei known as **isogametes**, morphologically they are indistinguishable. **Heterogametes** are designated as male and female gametangia and gametes that are morphologically different. Male gamete is known as **antheridium** and female gamete is known as **oogonium**.

**Nuclear cycle:** Like other living organisms, fungi have a cycle of haploid and diploid structure. The diploid phase begins with karyogamy and ends with meiosis. In the majority of fungi, however, there is no distinct alternation of haploid and diploid thalli.

**Heterokaryosis:** Nuclear cycle is not always clear cut as, it may appear from the foregoing statements. Nuclei of the same or of different genotypes may coexist side by side in the same mycelium and in the same cell of the hyphae. All cells do not necessarily have the same number of nuclei or the same kinds of nuclei; or the same proportion of each kind in a mixture of nuclei. The phenomenon of the existence of different kinds of nuclei in the same individual is known as **heterokaryosis**, and the individuals that exhibit this phenomenon are **heterokaryotic**.

**Parasexuality:** Some fungi do not go through a true sexual cycle, but derive many of the benefits of sexuality through **parasexuality**. This is a process in which plasmogamy, karyogamy and haploidization takes place, but not at specified points in the thallus or the life cycle. The deuteromycetes are fungi in which sexual reproduction does not take place and in which the parasexual cycle is of paramount importance. Some fungi that reproduce sexually also exhibit parasexuality.

**Sexual compatibility:** Compatibility is closely related to sex because, in a way, it governs sexual reproduction. Many fungi that produce clearly distinguished male and female sex organs on the same thallus but in which individuals are sexually self sterile because their male organs are incompatible, with their female organs and no plasmogamy can take place. Many other fungi do not produce differentiated sex organs at all. On the basis of sex, most fungi may be classified into following three categories:

(A) **Hermaphroditic:** In which each thallus bears both male and female organs that may or may not be compatible.

(B) **Dioecious:** In which some thalli bear only male and some thalli bear only female organs.

(C) **Sexually undifferentiated:** In which sexually functional structures are produced that are morphologically indistinguishable as male or female.

**Homothallic fungi:** Fungi in which every thallus is sexually self fertile and can therefore reproduce sexually by itself without the aid of mating types and non out crossing.

**Heterothallic fungi:** Fungi in which every thallus is self sterile and requires the aid of another compatible thallus of a different mating type for sexual reproduction. These fungi are out crossing.

**Secondary Homothallic fungi:** In this type of fungi during the spore formation two nuclei of opposite mating type are incorporated regularly into each spore or atleast some spores. Generally arising from those spores are self fertile and behave as they are homothallic, but in reality heterothallic. This condition is known as **secondary homothallism** or **Pseudohomothallism.**

# 2. SYSTEMATICS

For an organisms to be formally recognized by taxonomist, it must be described in accordance with internationally accepted rules and given a latin binomial name, a generic name followed by a specific name. Around 1964 about 64,000 fungal species were reported and since then about 1,500 species have been described and named as newly discovered each year. However, it is usually that about 60% of such discoveries are fungi that have already been found and described under a different name. Hence the number of fungi known is probably increasing at about 600 species a year, suggesting that about 70,000 species will be known in coming years. The last few decades have brought a number of changes in the study of fungal systematics and evolution. The major advances include – (1) recognition of the artificial nature of three or even five kingdom classification systems of organisms traditionally known as fungi (2) acceptance of the theory and data analysis techniques of phylogenetic systematics (3) development and application of molecular techniques in mycology and (4) additional discoveries of new taxa, including fossils. These developments signal a dynamic and exciting time for mycology.

Whittaker (1969) proposed a five kingdom classification of organigsms. The prokaryotes were accepted as constituting one kingdom, the monera, but the eukaryotes, in which there is a far greater number of species and structural diversity were divided into four groups on nutritional and structural basis. Unicellular eukaryotes (Protozoa and unicellular algae) were considered as a single kingdom, the protista. The multicellular eukaryotes, however, were subdivided on the basis of nutrition into three kingdoms, the photosynthetic plants (Plantae), the absorptive fungi (Fungi) and the ingestive animals (Animalia). Whittaker's scheme of classification is summarized in Fig. 2.1.

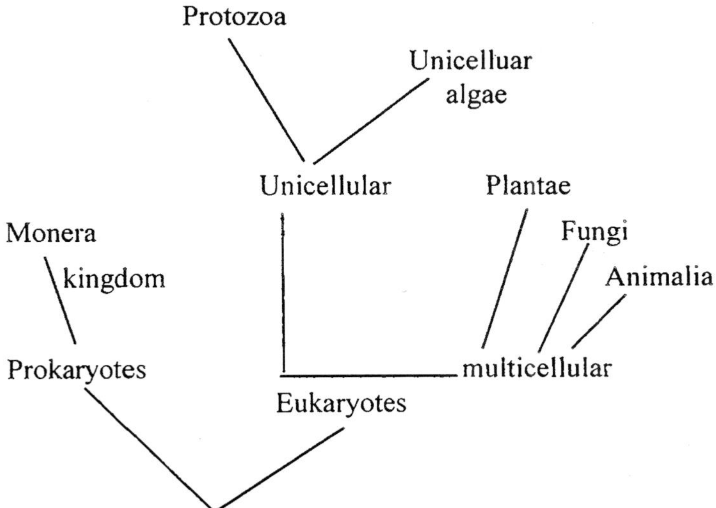


**Fig. 2.1:** Schematic representation of Whittaker's System of Classification

**Phylogeny:** It is a hypothesis of the genealogy of a group of organisms and their hypothetical ancestors. According to Alexopaulus and Mims (1996) a monophyletic group consists of an ancestor and all its descendants. Because Chytridiomycota, Zygomycota, Ascomycota and Basidiomycota form an inclusive group with the ancestors represented by the node that joins them, they are a **monophyletic** group included as fungi. **Polyphyletic** groups do not share a close common ancestor. The old concept of fungi including Oomycota and slime molds is clearly polyphyletic. The term **paraphyly** has been used in several different ways including to denote a group that includes some relatives of a common ancestor, but not all of them.

**Characters:** A character is any attribute or feature of an organism that can serve as a basis for comparison with other organisms. The different expressions of a character are known as **character states**. For example, fungal spores in a group under study may be smooth or ornamented. Each condition represents a different state of same character. Several characters are used to study their evolution including morphology, anatomy, ultrastructural features, biochemistry, nucleic acid secquences, and various other attributes. Anatomical characters of fungi can be observed with a compound microscope in squash mounts, hand cut sections or microtome sections of embedded material. Some of the anatomical characters used in mycology include the arrangement of hyphae comprising the tissues of spore producing structures, the arrangement of asci or basidia and sterile structures with in spore producing structures, and analysis of hyphal structure (Stevens, 1974).

A number of biochemical characters are significant in showing similarities between animals and fungi (Towe, 1981). Chemical techniques such as chromatography and protein electrophoresis have been useful in the comparison of fungal pigments and isozymes. The early use of, chemical spot tests in the study of lichenized fungi has developed, into every day use of chromatography to determine the presence of secondary compounds in the group. Other characters used in classifying fungi include decay type, especially in wood decaying species of basidiomycota, and the host relations of plant pathogenic fungi. Both of these characters are indicative of complex, readily tend themselves to phylogenetic analysis and can be polarized to provide information on the direction of evolution. A variety of genes have been sequences for fungi, including mitochondrial and several protein coding genes. Many phylogenetic sequencing studies of fungi have focussed on nuclear enclosed ribosomal RNA genes partly from practical considerations including the high copy number, unusual base pair ratio and early availability of primers for DNA (Lundberg and MC-Dade, 1950). It is also significant that the various regions with in RNA genes evolve at markedly different rates and provide resolution at a variety of taxonomic levels (Bruns *et al.*, 1991). Because rDNA has been studies in many incomplete fungi, large comparative data bases are now available.

**Species concept:** A morphological species concept is based on morphological characters alone. The systematists group individuals on the basis of observed similarities and distinguishes them from others on the basis of discontinuities in the characters. The biological species concept (May, 1942, 1963) defines a species as a natural population or populations of individuals that are actually or potentially interbreeding and are isolated reproductively from other such populations.

The Kingdom Fungi is divided into two kingdoms the Myxomycota and the Eumycota:

**I. Division-Myxomycota-The slime molds.**

Phylogenetically diverse organisms with amoebae as the trophic phase and also in some classes, plasmodia. Commonly known as slime moulds or slime fungi. They are distinguished from the true fungi in having a slimy mass of naked protoplasm without any cell wall. Most slime molds are saprophytic and some are parasitic. This division generally consists of three classes *i.e.*

(1) Class—Dictyosteliomycetes, cellular slime moulds. *Dictyostelium discoideum.*

(2) Class—Myxomycetes, plasmodial slime moulds. *Didymium iridis, Physarum polycephalum.*

(3) Class—Plasmodiophoromycetes. Obligate parasites of plants having minute intercellular plasmodia *Plasmodiophora brassicae.*

**Division Eumycota**

These are true fungi and comprises about 1,00,000 species. They are heterotrophic, eukaryotic organisms and differ from myxomycota in lacking a plasmodial stages and phagotrophic nutrition. Variation in fungal form is from unicellular form (yeast) to the fungi compared of mass of filaments, called hyphae. In majority of fungi cell wall is composed of a complex nitrogenous substance called chitin. Some of the true fungi have cell walls composed of chitin and cellulose. Majority of true fungi is eucarpic exception *Synchytrium*. The cell of the true fungi contain nuceli, nucleoli with a nuclear membrane. This division is divided into many subdivisions *i.e.* Chytridiomycota, Oomycota, Zygomycota, Ascomycota, Basidiomycota and Deuteromycota classes *i.e.* Chytridiomycetes, Oomycetes, Zygomycetes, Ascomycetes, Basidiomycetes and Deuteromycetes.

**Subdivision I: Chytridiomycota**

It contains single class chytridiomycetes. They produce motile cells at some stage in their life history. Each organisms posses a single, posteriorly directed, whiplash flagellum. They have cenocytic strcuture, an elongated simple hypha, well developed mycelium. Conversion of zygote into a resting spore or resting sporangium. Cell walls of these fungi contain chitin and glucon. Nuclear division is intranuclear and centric. There appear to be about 100 genera consisting of appoximately 1,000 species, most of which are saprobes. These fungi are pathogenic to plant, animal and fungal pathogens. Plant host includes vascular plants, mosses and phytoplankton. Animal hosts include nematodes, rotifers, mosquitoes, midges and some beetles. Fungal host include VAM fungi, Ascomycota and Basidiomycota. Chytrids are present in both aquatic habitats and soils. Few species are anaerobic species some of which have been shown to exist in the gut of herbivores belonging to a variety of mammalian families. Acquatic species that line mostly in the lower intermediate salinities found in estuaries. Most species complete their life cycles in few hours.

**Subdivision II: Oomycota**

Oomycota includes both unicellular, holocarpic forms and eucarpic, filamentous species composed of profusely branched, coenocytic hyphae that are often rather large and coarse in appearance. Septa are absent except at the bases of reproductive organs. Hyphae of species that attack higher plants grow in either an intracellular or an intercellular fashion. Obligate parasites tend to remain between host cells, giving rise to specialized hyphal branches termed haustoria that invade through the host cell wall and invaginate the host cell plasma membrane. Oomycetes are a common group of organisms found all over the world in both fresh and marine water as well as in terrestrial environments. Most of the aquatic forms-commonly referred to as water molds-grow in fresh water primarily, in well aerated streams, rivers, ponds and lakes. They are common in shallow waters near the bank or shore. Stagnant water sometimes exhibit distinctive communities of oomycetes. Most of the aquatic forms are saprobic on the remains of dead plants and animals. They play a major role in the degradation and recycling of nutrients in aquatic ecosystems. Some parasitic species that attack either algae or assorted animals such as nematodes, mosquito larvae and even carry fish. Majority of terrestrial oomycetes are either facultative or highly specialized parasites of vascular plants and cause some of the most serious diseases of certain agronomically important crops. *i.e.* Late blight of potato, root and fruit rots of citrus, downy mildew diseases of cucurbits, grapes, onions and lettuce, white rust of cabbage and damping off seedlings. Cell wall of Oomycetes is composed of β 1–3 and β 1–6 glucans and cellulose rather than chitin. *Achlya* and *Saprolegnia* are two common genera of Oomycetes. Asexual reproduction takes place by the formation of zoospores in zooporangium. Sexual reproduction takes place by the formation of oospore. Common genera are *Albugo*, *Bremia*, *Perenospora*, *Pythium*, *Phytophthora* and *Plasmopara*.

**Subdivision III: Zygomycota**

Zygomycota consists of two classes Zygomycetes and Trichomycetes. Trichomycetes is a group of fungi that are ecologically and morphologically distinct from all other fungi. Zygomycetes is the production of a thick walled resting spore called a **zygospore**. Zygospore develops with in a zygosporangium, that is formed by the fusion of gametangia. These fungi include a presence of a coenocytic mycelium asexual reproduction usually by sporangio-spores and absence of flagellate cells and centrioles. They may be homothallic or heterothallic. Mycelium is consists of coenocytic hyphae, the walls of which are composed of chitin, chitosan and polyglacturonic acid.

Asexual reproduction in zygomycetes may be by means of sporangiospores. Some species may produce conidia, chlamydospores, oidia and arthrospores. These fungi lack enzymes to degrade complex carbohydrates known as "**Sugar fungi**". Common species are, *Mucor, Rhizopus, Syncephalastrum, Cunnighamella, Gilbertella, Thaminostylum, Helicostylum, Circinella, Choanephora, Absidia* and *Syzygites*. Many species are capable of synthesizing important industrial products such as amylases, rennins and organic acids.

VAM fungi are kept in two orders–Endogonales and Glomales. Endogonales has one family with two genera. VAM/AM are formed by members of Glomales. Glomales consist of two sub orders--Glomineae and Gigasporineae. The former has two families with four genera and the later has one family with two genera.

Taxonomy of any group of plants is an important aspect for a botanist (Bursdall, 1990). During the last two decades systematics of VAM fungi has gained significance because of their role in soil fertility, nutrient uptake and biocontrol of plant diseases. Many of the fungi have not yet been cultured axenically which also includes VAM fungi. Therefore, their identification depends on specimens directly isolated from soil on the maximum observable characters.

Bursdall (1990) has also mentioned that "it is true that taxonomy still is based to a degree on opinion and suspect there is still too much", quick and dirty "taxonomy being published".

It is generally believed that VAM fungi evolved early in the history of vascular plants (Trappe, 1987). Inspite of their geological age (Berch, 1986) and their crucial role in origin of plants very little is known about their phylogenetic origin (Sancholle and Dalpe, 1993). The members of Glomales are believed to have been present as early as to Cambarian period (Pirozynski and Dalpe, 1989). They played a pivotal role in the origin of the terrestrial flora. Early Devonian plants showed the presence of the endosymbiont represented by non septate mycelia, coiled hyphae, irregularly shaped thin walled spherical structures resembling the vesicles (Remy *et al.*, 1994). Arbuscules like structures have recently been reported from plants preserved in Rhynie chert (Remy *et al.*, 1994).

Benjamin (1979) kept eight genera of endogonaceous fungi under Endogonales in a single family Endogonaceae. Morton and Benny (1990) divided this order in two independent orders Endogonales and Glomales on the basis of their spore structure and mycorrhiza development. The present Endogonales has one family, Endogonaceae with two genera, *Endogone* and *Sclerogone*. These fungi generally develop only arbuscules in plants from temperate parts of the world but most of the studies from tropical plants definitely report occurrences of vesicles in the root. Therefore, it will not be scientific to call these as AM fungi rather than the old terminology of vesicular arbuscular mycorrhizal (VAM) should be retained.

Taxonomy of Glomales is based on the structure of their spores/sporocarps (Morton, 1990; Hall, 1983). Some authors lay more emphasis on the wall structure which also is a valid criterion (Walker, 1983, 1992). Glomales are divided into two sub orders. Glomineae and Gigasporineae. The six genera included in this order are places under three families (Morton and Benny, 1990). Till present about 164 species of VAM fungi are genuine and validly published. The accumulation of data on indentification of VAM fungi has been scientifically done by Schenck and Perez (1990).

In the members of the family Glomaceae *i.e. Glomus* and *Sclerocystis* the spores are formed in sporocarps, spores arising in an orderly manner from sterile central plexus (Gerdemann and Trappe, 1974). Almeida and Schenck (1990) recognise only *S. coremioides* as valid species of *Sclerocystis*. The remaining species have been considered as synonyms of existing species of *Glomus*.

In *Acaulospora* and *Entrophosphora* belonging to family Acaulosporaceae, a saccule is formed terminally on a sporogenous hyphae, after which the spores are formed laterally or in between.

The spores in *Gigaspora* and *Scutellospora* of family Gigasporaceae are expanded from and borne on a bulbous sporogenous cell and have much larger spores than in other four genera (Spain *et al.*, 1990). Lot of literature exist on the taxonomy of Glomales. Most members of the sub order Glomineae develop vesicular arbuscular mycorrhiza and members of Gigasporineae from arbuscular mycorrhiza. The development of the two types of mycorrhiza and the distinction between different VAM fungi.

Of the total of about 164 VAM fungi known, 111 have been reported from India (Mukerji, 1996; Mukerji and Kapoor, 1986; Manoharachary *et al.*, 2002). 61 + 9 + 20 + 15 + 3 + 2 + 1 = 111 These fungi are represented by 61 species of *Glomus*, 9 sp. of *Gigaspora*, 20 sp. of *Acaulospora*, 15 sp. of *Scutellospora*, 3 sp. of *Entrophosphora*, 2 sp. of *Endogone* and 1 of *Sclerocystis*.

**Subdivision IV: Ascomycota**

Distinguishing factor of these members of Ascomycota from all other fungi is the *ascus* a sac like cell containing the ascopores. Eight ascospores are typically formed with in the ascus, but this number may vary from one to over a thousand according to the species. Septa having simple pores and the presence of woronin bodies. Concentric bodies are present in the lichen forming ascomycetes. Eukaryotic organism in their ability to, occupy a broad range of habitats. Some species can be found in their habitats most of the year. Plant parasitic ascomycetes may develop ascospores only after the death of their hosts and their reproductive structures are found in the early spring when the spores are mature and about to be released. Some of the fungi are coprophillous, while others grow on bark, wood and leaves known as corticolous, lignicolous or folicolous. Ascomycetes are of great importance in human affairs. The cellulolytic ascomycetes such as *Chaetomium* and *Trichoderma* are among the organisms responsible for the destruction of cellulosic fabrics. As oomycetes are among the worst enemies of human beings.

*Sporothrix, Pneumocystis carinii* the casual agent of a virulent pneumonia that attacks individuals with suppressed immune systems. Ascomycetes are closely associated with insects. Some such as *Ophiostoma, Ambrosiella, Raffaelea* and *Symbiotaphrina* are found in insects structure and provide or detoxify the food of the insects. Laboulbeniales are biotrophic parasites of insects and few other arthropods. *Beauveria, Metarrhizium* and *Fusarium* kill insects under appropriate environmental conditions and are being studied for use as biological insecticides.

Ascomycetes are single celled, mycelial or diomorphic. A large proportion of the cell walls of filamentous ascomycetes is of chitin. Cellulose also has been reported in the hyphal walls of several unrelated ascomycetes. In Ascomycetes wall appears to be two layered. Thick, translucent inner layer and a dense, thin outer layer. Hyphae are divided into compartments by septa that form from the hyphal periphery and advance towards the centre. Asexual reproduction takes place by fission. Fragmentation or formation of chlamydospores or conidia are according to species and environmental conditions. Antheridia and Ascogonia are present in the sexual reproduction. *Neurospora crassa, Pyronema, Aspergillus, Penicillium, Xylaria, Protomyces, Taphrina, Saccharomyces, Chaetomium* and *Ascobolus* are common genera.

**Subdivision V: Basidiomycota**

Basidiomycetes are large and include forms commonly mushrooms, puffballs, earthstars, stinkhorns, bird's net fungi, jelly fungi, bracket or shelf fungi, rust and smut fungi. Basidiomycetes are characterized by the fact that they produce their sexual spores termed basidiospores, on the outside of a specialised, microscopic, spore producing structure called the basidium. Basidiospores are formed after plasmogamy, karyogamy and meiosis. Four basidiospores are formed on the basidium.

Basidiomycetes are important group of fungi including harmful as well as useful species. Notorious among these diseases are stinking smut and black stem rust of wheat. Many others attack a large variety of food and ornamental plants. These fungi have a well developed septate hyphae, that grow through the substrate and absorb nourishment. Hyphae are microscopic, mycelium is usually white, bright yellow or orange and often spreads out in a fan shaped growing front. **Rhizomorphs** commonly are produced by many ectomycorrhizal species and by various wood rotters and are important not only in the spread of certain species. Mycelium is heterothallic, primary mycelium, secondary mycelium and tertiary mycelium. Basidiomycetes produce their basidia in fruiting bodies of various types of basidiocarp. Basidia are typically formed in definite layer called **hymenia**. Aecia, teliospore, uredinia and basidia are present in Basidiomycetes.

Mycorrhiza, *Coprinus*, Mushrooms, *Geastrum, Ganoderma, Polyphora, Polystictus*, *Hydnum*, rust and smut fungi.

**Subdivision VI: Deuteromycota**

Fungi which never reproduce sexually have been referred to as deuteromycetes or imperfect fungi to denote what was considered to be their status as second class members among perfect sexually producing fungi. Asexual ascomycetes and basidiomycetes have recognized as deuteromycetes. Vast majority of the fungi are terrestrial, although a number of species do occur in marine and fresh water habitats. Majority of deuteromycetes are either saprobes or weak parasites of plants. Few species even trap and consume nematodes known as predacious fungi. A number are parasitic on other fungi and lichens. Many conidial fungi are of great importance because, they are parasites and cause serious diseases of plants and animals including humans. Some are used in the commercial production of certain chemical including some antibiotics. These fungi have septate, well developed branched hyphae. Conidiophores may be formed singly or grouped together to form specialized structures conidiomata, synnemata, sporodochia, or produced in conidiomata known as pycnidia or acervuli. In **synnemata** the stalk like portion is longer in comparison to the branched top and the entire structure resembles a long handled feather cluster. In **sporodochium**, the conidiophores arise from a central cushion shaped pseudoparenchymatous stroma. Conidiophores are packed closely together and generally are shorter than those composing a synnema.

Pycnidium is a globose, flask shaped pseudoparenchymatous structure that is lined on the inside with conidiophores. Some Pycnidia resemble perithecia. **Pycnidia** may be completely closed or may have an opening (ostiole). An **acervulus** is typically a flat or saucer shaped bed of short conidiophores growing side by side and arising from a more or less stromatic mass of hyphae. Acervuli typically are produced in plant tissue subepidermally or subcuticularly and eventually break through the plant material to become erumpent. Conidium is nonmolite, asexual spore formed at the tip or side of a sporogenous cell. Macroconidia and microconidia are present. Commonly occurring genera are *Cladosporium, Humicola, Penicillium, Alternaria, Curvularia, Drechslera, Colletotrichum, Torula, Fusarium, Trichoderma, Verticillium, Acremonium, Paecilomyces* etc.

Classification of fungi According to Ainswonth and Bisby's is described here which based on the thallus morphology, Unicellular hyphae, multicellular hyphae, reproductive parts and zoospores.

According to Ainsworth Bisby's (1971)

**Divisions:**

I. **Myxomycota** (Myxobionta). Somatic phase amoebae, aggregation of amoebae or a plasmodium.

**Division: Myxomycota:**

(1) **Classes Acrasiomycetes:** assimilative phase free living, amoebae which unite as as pseudoplasmodium before reproduction.

(2) **Hydromyxomycetes:** assimilative phase amoebae, pseudoplasmodium or a net plasmodium.

(3) **Myxomycetes:** assimilative phase free living, saprobic plasmodium.

(4) **Plasmodiophoromycetes:** plasmodium parasitic with cells of host plant.

II. **Eumycota** (Mycobionta, eumycetes)

Somatic phase not amoeboid or plasmodial, typically mycelial but some times unicellular.

**Subdivisions:**

(1) **Mastigomycotina:** Thallus unicellular or mycelial, spores and gametes motile.

(2) **Zygomycotina:** Thallus mycelial and typically aseptate; sexual reproduction results in a resulting spore (zygospore); motile cells absent.

(3) **Ascomycotina:** Thallus mycelial and septate or rarely, unicellular, sexual reproduction by ascospores developed endogenously in asci.

(4) **Basidiomycotina:** Thallus mycelial and septate or clamp connections are characteristics, sexual reproduction by basidiospores borne exogenously on basidia.

(5) **Deuteromycotina:** Thallus mycelial and septate or unicellular, sexual reproduction absent but parasexual cycle may occur.

**Subdivision:** Mastigomycotina

**Classes**

(1) **Chytridiomycetes:** Zoospores has 1 posterior whiplash flagellum.

(2) **Hypochytridiomycetes:** Zoospore has 1 anterior tinsel flagellum.

(3) **Oomycetes:** Zoospore has 2 equal flagella, a tinsel flagellum directed forward and a whiplash flagellum directed backward.

**Subdivision: Zygomycotina**

**Classes**

Zygomycetes—cosmopolitan, saprobes and parasites.

Trichomycetes—parasite on arthropods, thallus branched or simple.

**Subdivision: Ascomycotina**

**Classes**

Ascomycetes—ascospores are developed by free cell formation.

**Subclass**

Hemiascomycetes—ascocarp absent.

Plectomycetes—ascocarp typically a cleistothecium

Pyrenomycetes—ascocarp a perithecium.

Discomycetes—ascocarp an apothecium.

Laboulbeniomycetes—ectoparasites of arthropods.

Loculascomycetes—ascocarp an ascostroma.

**Subdivision: Basidiomycotina**

Basidium bearing basidiospores is present, clamp connection is present. Three classes may be distinguished according to the presence or absence of a well developed basidiocrap.

**Class 1: Teliomycetes:** Basidiocarp replaced by a sorus producing resting spores which function as probasidia.

**Class 2: Hymenomycetes:** Basidiocarp usually well developed and typically gymnocarpous or semiangiocarpous, basidiospores ballistospores.

**Class 3: Gasteromycetes:** Basidiocarp well developed, typically angiocarpous, basidiospores not ballistospore.

**Teliomycetes:** Uredinales—Teliospore is terminal and germinates to give a promycelium bearing basidiospores on sterigmata from which they are forcibly discharged.

Ustilaginales—Teliospore are intercalary or terminal or laterally produced basidiospores being sessile and not forcibly discharged from the metabasidium.

**Hymenomycetes:** Phragmobasidiomycetidae (Heterobasidiomycetes) in which the metabasidium is divided by primary septas, usually cruciate or horizontal.

Order —Tremellales
—Auriculariales
—Septobasidiales

Holobasidiomycetidae—in which the metabasidium is not divided by primary septas but may sometimes become adventitiously septate.

Order —Exobasidiales
—Brachybasidiales
—Dacrymycetales
—Tulasnellales
—Aphyllophorales
—Agaricales

**Deuteromycotina:** Three classes

Balastomycetes—yeasts and yeast like forms.

Hyphomycetes—mycelial forms which are sterile or bear conidia on hyphae or aggregations of hyphae but not in discrete sporocarps.

Coelomycetes—forms producing conidia in pycnidia or acervuli.

According to the new classification of Alexopoulous et. al., 1996, Division of Fungi has been changed into Phyla—which includes the Phyla-Chytridiomycota, Zygomycota, Ascomycota and Basidiomycota. Additional Phyla that are considered include—Myxomycota, Dictycosteliomycota, Acrasiomycota and Plasmodiophoromycota.

According to Alexopoulous, Mims and Blackwell (1996)

Kingdom Fungi

Phylum —Chytridiomycota

Class —Chytridiomycetes

Order —Spizellomycetales—*Rozella, Olpidium*

—Chytridiales—*Synchytrium, Rhizophydium, Chytriomyces*

—Blastocladiales—*Allomyces, Blastocladiella, Coelomomyces, Catenaria, Physoderma*

—Monoblepharidales—*Monoblepharella*

Phylum —Zygomycota

Class —Zygomycetes

Order —Mucorales—*Mucor, Rhizopus, Gilbertella, Dicranophora, Saksenaea, Phycomyces, Absidia, Mortierella Pilobolus, Choanephora, Thamnidium Cunnighamella, Mycotypha, Synchephalastrum, Sigmoideomyces, Dimargaris*

—Kickxellales—*Coemansia* and *Kickxella*

—Endogonales—*Endogone, Sclerogone*

Order —Glomales—*Glomus* sp.

—Zoopagales— *Zoophagus, Acaulopage*
*Cystopage, Stylopage*
*Zoopage and Piptocephalis*

Class —Trichomycetes

Order —Asellariales —*Asellaria*
—Harpellales —*Harpella* and *Smittium*
—Eccrinales —*Eccrina*
—Amoebidiales —*Amoebidium, Paraamoebidium*

*Phylum* —Ascomycota

Class —Deuteromycetes (Asexual Ascomycetes)

Class —Archiascomycetes

Order —Taphrinales —*Taphrina* and *Protomyces*

—Schizosaccharomycetales—*Schizosaccharomyces*

—Saccharomycetales—*Saccharomyces*.

—Eurotiales —*Penicillium, Aspergillus*

—Hypocreales —*Nectria, Hypomyces, Hyporea* and *Claviceps*

—Microascales —*Microascus, Ceratocystis*

—Phyllachorales —*Glomerlla, Phyllachora* and *Polystigma*

—Ophiostomatales —*Diaporthe, Gnomonia*

—Xylariales —*Xylaria, Camillea, Daldinia* and *Rosellina*

—Sordariales —*Neurospora, Sordaria, Podospora*

—Meliolales —*Meliola*

Class —Discomycetes

Order —Medeolariales —*Medeolaria*

—Rhytismatales —*Rhytisma*

—Ostropales —*Graphis*

| | | |
|---|---|---|
| Order | —Cyttariales | —*Nothofagus* |
| | —Helotiales | —*Monilinia, Sclerotinia* |
| | —Gyalectales | —*Coenogonium, Treutepohlia* |
| | —Lecanorales | —*Peltigera, Lecanora, Parmelia* |
| | —Caliciales | —*Trentepohtlia, Chlorella* and *Stichoccoccus* |
| | —Pezizales | —*Peziza, Ascobolus, Pyronema* and *Morchella* |
| Class | —Local ascomycetes | |
| Order | —Coryneliales | —*Podocarpus, Corynelia* |
| | —Dothideales | —*Dothidea, Aureobasidium* |
| | —Arthoniales | —*Asterina* |
| | —Capnodiales | —*Capnodium, Chaetothyrium, Treubiomyces* and *Cephaleurous* |
| | —Patellariales | —*Rhytidhysteren* |
| | —Pleosorales | —*Pleaspora* |
| | —Erysiphales | —*Erysiphe, Sphaerotheca, Uncinula Microsphaera* |
| | —Laboulbeniales | —*Pyxidiophora, Mycorhynchidium* and *Laboulbenia* |
| | —Spathulosporales | —*Spathulospora* |
| Phylum | —Basidiomycota | |
| Class | —Basidiomycetes | |
| Order | —Agaricales | —*Agaricus, Tricholoma* etc. |
| Class | —Gasteromycetes | |
| Order | —Lycoperdales | —*Calvatia, Lycoperdon* etc. |
| | —Tulostomatales | —*Calostoma, Tulostoma* and *Battarrea* |
| | —Sclerodermatales | —*Scleroderma* and *Astraeus* |
| | —Phallales | —*Phallus* |

Order —Nidulariales —*Nidularia, Sphaerobolus*

—Aphyllophorales—*Phellinus, Ganoderma, Polyporus, Schizophyllum, Cantharellus, Clavaria* etc.

—Uredinales —*Puccinia, Phragmidium, Uromyces*, and *Melampsora*

—Ustilaginals —*Ustilago, Tilletia*

—Tremellales —*Tremella*

—Auriculariales —*Auricularia*

—Ceratobasidiales—*Ceratobasidium, Rhizoctonia*

—Tulasanellales —*Tulasanea*

—Sporidiales —*Rhodotorula*

—Septobasidiales —*Septobasidium*

—Exobasidiales —*Exobasidium*

Rhylum —Oomycota

Class —Oomycetes

Order —Saprolegniales —*Saprolegnia, Aphanomyces*

—Lagenidiales —*Lagenidium*

—Leptomitales —*Leptomitis, Plerogone*

—Rhipidiales —*Rhipidium*

—Perenosporales —*Perenospora, Albugo, Pythium Bremia* and *Plasmopara*

Phylum —Hyphochytriomycota

Class —Hyphochytriomycetes

Order —Hyphochytriales—*Rhizidiomyces*

Phylum —Labyrinthulomycota

Order —Labyrinthales

Genus —*Labyrinthula, Thraustochytrium*

Phylum —Plasmodiophoromycota

Class —Plasmodiophoromycetes

Order —Plasmodiophoromycetales—*Plasmodiophora*

Phylum —Dictysteliomycota

Class —Dictyosteliomycetes

—*Dictyostelium*

Phylum —Acrasiomycota

Class —Acrasiomycetes

—*Acrasis*

Phylum —Myxomycota

Class —Myxomycetes

Order —Liceales —*Lycogala, Tubifera, Dictydium,*

—Echinosteliales —*Echinostelium*

—Trichiales —*Trichia, Hemitrichia, Arcyria*

—Physarales —*Physarum*

—Stemonitales —*Stemonitis*

—Ceratiomyxales —*Ceratiomyxa*

# 3. ECOLOGY

Ecology is the 'study of the relations of animals and plants. Particularly of animal and plant communities, to their surrounding, animate and inanimate. A small volume of soil contains a vast assemblage of certain organisms of diverse form and physiology. Their small size, however precludes the morpholoigcal complexity attained by higher plants and animals so that their reactions to each other and to their environment should be more easily attributable to basic physiological processes than is usually the case.

In fungi, the vegetative hyphae approaches the simplicity and uniformity of structure of the bacteria cell but identification is more readily made by virtue of their differentiated reproductive stages. The bacterium unicellular, small and with a large surface to volume ratio is well suited for the absorption of nutrients from a solution in which the cell is increased. The hyphae of fungi and actinomycetes, however persuit the penetration of solid nutrient substrates by physical force and the extension of the mycelium from one substrate to another over nutritionally uncongenial areas.

**Soil:** Soil is the surface layer of the land, consisting of an mixture of inorganic and oganic components physically and chemically distinct from the underlying material. Most roots and soil organisms are located in the upper zone, thickness of which is measured in inches rather than feet.

**Soil Profile:** In vertical section soil have various layers. The top layers, largerly organic in content.

Soil Profile: In vertical section soil is divided into various layers. The top layers, largely organic in content. The highest horizon that is predominately mineral is designated A and is leached. It is frequently divided into two subhorizons. The $A_1$ horizon is distinguished by a relatively high organic matter content and usually dark colour. Beneath the $A_2$ horizon has lost clay minerals, iron or aluminium, or even all three, so there is a concentration of the more resistant minerals. Cultivation destroys the top of the profile and a mixed horizon. B horizon is characteristically illuvial, there is an accumulation of clay, iron, aluminium. Although roots and organisms penetrate into the $A_2$ and B horizons, maximum biological activity is in the $A_1$ hroizon.

In humid areas, leaching is intense and a marked A horizon develops, frequently acid in reaction. In drier climates however leaching is reduced and the A horizon may be shallow and neutral to alkaline. B horizons usually contain an accumulation of calcium carbonate and calcium sulphate, even though the parent rock may not be calcareous.

**Organic matter:** This organic layer is further divided into three layers L, F, and H. In the majority of soils, the surface layer consists of relatively unaltered mass of identifiable plant remains. L layer is litter layer or $A_x$ horizon. In a mor soil the F, (Fermentation) layer lies beneath the L and consists of plant remains in the process of decomposition. The lowest organic layer is the H (Humus) layer is which the organic material is amorphous and no longer identifiable. The F and H layers together form the $A_x$ horizon. In a **mull** soil the F and H layers are usually not distinguishable because the activity of soil animals, particularly earth worms, leads to a rapid intermixing of the decomposing organic material with the surface mineral layer of the soil. In a mull soil, the L layer thus rests virtually directly on the $A_1$ horizon. Intermediate conditions between mull and mor are frequent and are sometimes referred to as '**mor**'.

In mor soil the rate of decomposition is very slow. Phenolic materials in the plant tissues tan the protiens and also form protective coatings over cellulose fibrils thus rendering them relatively unavailable to microorganisms and the meiofauna. Such soils are also charcterized by low base status so that they tend to be very acid.

With in the A and B horizons, oganic matter can be divided into two components, are derived from the L, F and H layers.

Organic material formed *in situ* consists of the soil animals, large or small and the walls and cellular contents of bacteria, fungi and other microoganisms, but principally of the roots and other undergorund organs of the plants when dead, roots are clearly a part of the soil but they are not considered alive. Exudates from the roots, influence microbial activity with in the rhizosphere. The rhizoplane flora is largely saprophytic, organic matter from the aerial parts of plants becomes incorporated into the mineral layers of the soil, it has been greatly changed by a whole series of the biological activities. The humified orgaic matter may still contain some of the more resistant original polysaccharides but the main consists of mixtures of humic acid, fulvic acid and humin.

**Inorganic matter:** The inorganic particles are derived, directly or indirectly from parent rocks and their composition. Many of the rock minerals, however are changed by the physical and chemical processes of weathering and most of the smaller particles are secondary in nature. Soil is thus a complex product of parent material, topography, climate, time and biological activity. The major mineral particle that occurs in its primary form in soil is quartz and most of the larger particle fraction consist of this unreactive silica.

The size of the soil particles have a continuous distribution to divide them into fractions is arbitrary. The more important fractions of these, the International and American systems are given below Table 3.1.

**Table 3.1:** Name and size of main soil fractions according to the International and American Systems

| Name of Fraction | Equivalent diameter of particles (mm) | |
|---|---|---|
| | International | American |
| Sand | 2.0–0.02 | 2.0–0.05 |
| Silt | 0.02–0.002 | 0.005–0.002 |
| Clay | <0.002 | <0.002 |

The terms clay, loam and sand are thus arbitary divisons of a continuous scale of shear strength. Soils of intermediary shear strength are loams.

**Soil aggregates:** The individual particles of a soil are bound together into aggregates and the term structure refers to the frequency, size, shape and stability of these aggrretes. Soil structure profoundly affects the physcial and chemcial properties of the soil. In terms of structure, soils may be platy, prisonatic, blocky or spheroidal and there are many subdivisions for ex-columnar, angular and granular. Emerson (1959) formulated a model of the structure of soil aggregates, clay crystals are grouped together into packs, or domains oreinted platelets. Different domains may be held together by electrostatic attraction but in the main, the binding substance is thought to be organic matter, binding quartz to quartz, quartz to clay domain and domain to domain. In the pores of soil aggregates, small bacteria and fungi cannot enter and much of the organic matter in a soil can be protected from degradation. The soil pores are filled in various proportions with an aqueous solution and gases of which oxygen, nitrogen and carbon dioxide are the most important.

**Nutrition of Fungi:** Fungi are heterotrophic in that they are able to synthesize organic material sufficient for growth from performed organic components. They are also chemotrophic, they obtain energy independently of light by chemical reactions. The energy yielding reactions of fungi are based on organic substances and respiration is typically aerobic, little growth occuring in the absence of oxygen as a terminal electron acceptor. Fungi in culture are able to utilize the common hexoses and disaccharides for a carbohydrate in the external medium to act as a respiratory substrate, it must be able to pass through the cell wall and the cell membrane. Entrance of sugars is effected by carriers mechamisms (sometimes called permeases) specific to each sugar, possibly specific to groups of sugars with the same molecular shape.

The *Saccharomyces cerevisiae* possesses a carrier for maltose which is thus able to enter the cell where the α-linkage is cleaved by maltase to yield two glucose molecules that are subsequently phosphorylated. *S. cerevisiae* lacks however a carrier for sucrose and this sugar is cleaved extracellulary to glucose and fructose by invertase located on the cell surface. Carriers are sometimes inductive systems, being formed only in the presence of the specific sugar group and inhibition can also occur.

The greater quantity of respiratory substrates occurring naturally in soil are derived from the cell wall of plants which consist primarily of complexes of sugars or sugar containing molecules. The degradation of these complexes involves an extraordinary number of enzymes. Exo-β-1-3-glucanase, has little ability to degrade β 1-4 or β 1-6 glucans, while a β-1-4 Glucanase does not hydrolyze β-1,4 glucans or β-1,4 galactans.

Carbohydrate utilization is further complicated by the phenomena of induction, inhibition and supression. Some enzymes, like carrier systems, are produced only in the presence of the substrate and are then termed inductive. Cellulase synthesis in *Pyrenochaeta terrestris* is subject to induction but it is also repressed by glucose in concentration above 0.0005 molar. Carbohydrate utilization, can be blocked at many points. Entrance of simple carbohydrates may not occur because of the lack of the relevant carrier system or it may not be induced or it may be inhibited. Even if the carbohydrate enters the cytoplasm, it may not be utilized because of the absence of the necessary phosphorylating enzyme.

**Nitrogen:** Aminoacids are related compounds are present in small quantites in soil, particularly in the rhizosphere, most nitrogen available for uptake is in the form of ammonia or nitrate. Most fungi are able to utilize both ammonia and nitrate although ammonia is prefered (Nicholas 1965). Many soil fungi, however appear to need amino acids or yeast extract in order to grow their maximum rate (Atkinson and Robinson, 1955). Many wood destroying basidiomycetes too, fail to use nitrate but grow luxuriantly on amino sources. Nitrate utilization is dependent upon the induction of nitrate reductase and is inhibited by ammonia, and ammonia is largely utilized before nitrate.

## Nutrition and Substrate Colonization

**Successions:** The breakdown of natural products by the m[illegible]ed soil flora and fauna is such a complex process. Tribe (1960), however studied the colonization and breakdown of cellophane film and his work provides an insight into the factor involved.

Breakdwon of cellophane buried in a number of soils was studied and although there were variations in detail from soil to soil.

The first colonizers were fungi and three groups based on vegetative morphological characters. The members of the first group were characterized by coarse mycelium that ramified over the surface of the cellophane and rapidly initiated decomposition by lysis of the cellulose adjacent to their hyphae. Such fungi, mostly in the form of genus *Rhizoctonia*, were clearly strongly celluloltic and were dominant organisms on the substrate at this stage. In some soils, members of a second group, often species of the well known cellulolytic genera *Humicola, Botryotrichum* and *Chaetomium* appeared as codominants. These fungi didnot ramify extensively in contact with the cellophane but rather penetrated its thickness at scattered sites by means of rooting hyphae which then branched out to form an approximately circular hyphae system within the thickness of the cellophane. The third group of initial colonizers. Consisted of Chytrids that commonly developed on some cellophane pieces.

The presence and activity of these initial colonizers were related to their ability to produce cellulases. Other factors, however, must have been involved because the soils contained many other cellulolytic fungi which rarely appeared on the cellophane. The extent of lysis produced by the various fungi obviously depended upon physiological and morphological factors.

Bacteria were relatively uncommon during this initial fungal phase and appeared to cause little lysis of the cellophane. During mycelial senesence, however, they rapidly increased and presumably utilized either materials diffusing from the hyphae or the hyphae themselves. Carbohydrate is a key factor in ecology and that the diverse chemical substrates in soil (Gupta, 1967) will support an equally diverse microbial population and zygomyetes but the representatives in other classes. These fungi are characterized by an inability to utilize anything but the simplest carbohydrates usually sugars, including starch. Following the colonization of the complex substrate by sugar fungi, it might be supposed that there would be successive waves of colonization by fungi that utilize progressively more refractory components. Lignin decomposers utilizing a substrate that relatively few organisms can degrade, and slow growth rate might well be imposed upon them by the slow rate of enzymic actions breaking down the lignin to the basic monomers.

This biochemically based succession has been thought to be reflected in a taxonomic one, as the Phycomycetes have been considered to be sugar fungi par excellence, the Ascomycetes and Fungi Imperfecti to be frequently cellulolytic, and lignin decomposition to be primarily the work of the higher Basidiomycetes (Garrett, 1951). Most fungi imperfecti are important in the degradation of lignin, tannin and related compounds (Domsch, 1960; Lewis and Starkey, 1969).

If a dung is studied for the succession of fungi. The first fungi appear to be are zygomycetes, followed by ascomycetes and basidiomycetes, and in general terms these have been considered to represent sugar, cellulose and lignin utilizers respectively. Harper and Webster (1964) confirms the sequence, have shown that it is not a succession, based on nutritional factors.

## MICROBIAL INTERACTIONS

### Types of Microbial interactions

**Commensal and mutualistic interactions:** In natural habitats, microorganisms of one species usually grow in close association with those of many other species. Sometimes the association may be without discernible mutual influence and is then termed 'commensal'. Mycorrhiza are well known exmaples of mutualistic symbiosis where both partners appear to benefit from the association.

**Antagonistic interactions:** Importance of antagonism is furnished by the difficulty of establishing a fungus in a natural soil in which it does not normally occur. The most famous example of antagonism is ascomycete *Ophiobolus graminis*. This fungus causes take all disease of wheat but also attacks many other monocotyledons, usually invading their roots. Although host infection occurs from hyphae lying in the infected remains of previous crops, infection by ascospores is rarely successful.

Take all disease gradually decreases after a few years if the site is sown annually to wheat. This unexpected phenomenon of take all decline appears to be attributable to increased antagonism to *Ophiobolus graminis* during both its saprophytic and parasitic phases. Antagonism, may be attributed to exploitation, competition or antibiosis. Exploitation implies the utilization of the cells of one organism by another, the exploited cells being either living or dead, but if the latter, killed by the exploiter.

**Exploitation and lysis:** Some soil animals, particularly mites feed largely on fungal spores and that many fungal structure are ingested by worms. Nematodes too pierce fungal walls and extract the protoplasm. Fungi exploit other fungi, interaction usually being termed parasitism. Although biotrophic relationship are known for extraterrestrial mycoparasites, and for a few with in soil, a cruder necrotrophic relationship predominates among soil fungi, where the explotter in one interaction may be the exploited in another. Mycoparasitic capabilities have been demonstrated for such common soil fungi as *Trichoderma viride, Rhizoctonia solani, Trichothecium roseum* and *Penicillium* sp.

Bacteria can penetrate the walls of fungi and multiply with in them (Old, 1969), and it is likely that soil Streptomycetes can do the same. However cell wall lysis and disentegration of the cytoplasm occur while the bacterium or Streptomycetes is still outside the wall.

**Fungistasis, antibiosis and competition:** In soil some fungi are inactive, existing manily in the form of spores but sometime are resting hyphae. This exogenous dormancy of much of the soil population has been termed 'fungistasis'. Fungistasis is widespread in natural soils, is non-specific, disappears in the absence of microorganisms, and is temporarily annulled by the addition of nutrients. Subsoils are rarely fungistatic. Fungistasis can be restored to a sterile soil by mixing in a little unsterile soil.

Non sterile soil extracts and diffusates are often fungistatic and this was originally thought to imply the presence of water soluble inhibitory chemicals produced as the result of microbial metabolism. These metabolites become deleterious, inhibit germination and cause morphological changes in hyphae (Park, 1961). Bacteria on or near the spore surface may even cause such a negative concentration gradient that spores lose nutrients, thus accounting for fungistasis in fungi reported to germinate solely on endogenous nutrients in pure culture.

Antibiosis is best considered in two sections, related to the two concepts of an antibiotic. Compounds are usually produced in greatest concentrations, when there are ample nutrients, a condition not to be found in the bulk of the soil. Generally distributed specific antibiotic can be found in heavily amended soil, an example being provided by patulin produced by *Penicillium urticae* in stubble mulched soil—(Norstadt and MC Calla, 1969). Gliotoxin has been demonstrated in high concentration in straw and seed coats in an acid soil which naturally contained the producing organisms. There is thus evidence for the occurrence of specific antibiotics in pieces of natural substrate that provide sites with a relatively high nutrient status. It is in such sites, where active colonization is occurring, the production of inhibitors might favour the producer in its competition with other organisms. The production of specific antibiotics in such localized sites is probably wide spread but needs confirmation.

# 4. PHYSIOLOGY

An animal cell typically consists of a nucleus with associated cytoplasm bounded by a cell membrane. Plant cells have an external cell wall to the membrane. Yeasts approximately spherical in form and with a single nucleus, recognized as a unicellular organisms. Ultrastructure studies of filamentous fungi demonstrate a typical eukaryotic cellular organization.

The hyphae of a single species can differ considerably in diameter depending on environmental conditions and their position in a colony. Hyphae at the margin of a mature colony of *Aspergillus nidulans* are about 3-4 μm wide, where as comparable hyphae in *Neurospora crassa* can have a diameter of 10 μm or more. The hyphae of these and other fungi are divided into compartments by cross walls. In *A. nidulans* the apical compartment including the hyphal tip commonly has a length of 300-400 μm and subsequent compartments an average length of about 50 μm. The hyphae of most lower fungi are not subdivided into compartments, it may appear hemispherical in shape.

The hyphal compartments of many fungi may contain one or many nuclei. In Ascomycetes number of nuclei can vary from 4-50 in number. While in the case of Basidiomycetes the monokaryon has one nucleus per compartment and the dikaryon two, although some basidiomycetes have many nuclei per compartment. Yeasts have a single nucleus per cell. The nuclei of the vegetative phase of most filamentous fungi are haploid, while in oomycetes it is diploid. Chromosomes of fungi contain an amount of DNA similar to that of prokaryotic chromosome and the haploid nucleus an amount only a few times greater than the prokaryote cell. Nucleosomes occur in the base pairs of DNA wrapped around a disc consisting of four different histones. Main difference between fungal and eukaryotic nucleosomes are in the linking DNA between nucleosomes, which is shorter than the 60 base pair link of higher organisms and in the histone which is associated with the linking DNA, which throughout the eukaryotes is more variable than the disc histones.

Mitochondria are the site of oxidative phosphorylation in eukaryotic cell. The enzymes concerned with electron transport and ATP Production are located on the inner membrane of the mitochondria, area increased by invagination is cristae. Mitochondria can vary in size form and numbers during the cell cycle and in response to environmental conditions. Cells of *Saccharomyces cerevisiae* may contain one or a few much branched mitochondria or many up to about twenty, small unbranched mitochondria. Fusion occur between mitochondria, and fragmentation of a large mitochondrion can take place to yield several smaller ones.

Mitochondria contain DNA, which may form a nucleoid at the centre of the mitochondrial DNA molecules, which can be extracted and examined by electron microscopy. An mt DNA molecules normally consists of a circle of double stranded DNA, but in a few microorganisms including the yeast *Hansenula mrakii*, the molecule is linear.

**Plasmids:** Plasmids are pieces of DNA that are capable of replication independently of the replication of the chromosomes or mitochondrial DNA. Although plasmids are very common in bacteria, they are in fungi and other eukaryotes. An intensively studied fungal plasmid is 2 μm DNA that occurs in the nucleus of *S. cerevisiae*. The 2 μm circular double stranded DNA molecule consists of about 6200 base pairs. The plasmid can account for about 3% of the DNA is a cell and there can be upto 100 copies per cell. Whether the 2 μm DNA has any role in the life of the host is unknown, but it providing valuable role in the construction of vectors for the genetic manipulation of yeasts. Another yeast *Kluyveromyces fragilis* has a plasmid in the form of two linear ds DNA molecules of different lengths. Strains with the plasmid are killers producing an extra cellular protein that kills *K. lactis* strains that lack the plasmid. In *Podospora anserina* plasmids that originate from mitochondrial DNA can spread through the mycelium and cause senescence in which growth rate falls and vegetative propagation finally becomes impossible. The plasmids can also be transmitted via the maternal parents through ascospores. Senescence in the progeny may be long delayed, allowing the continued propagation of the plasmids. It would seem that fungal plasmids are essentially 'Selfish DNA' concerned with their own propagation.

The object of the study of fungal physiology is to determine as far as possible how the living fungus works. A better understanding of the relation between the fungus and its environment is an aid to this end. The actual environment of most fungi is however so complex that it is almost impossible to trace the effect of any one factor. In most physiological experiments with fungi the organism is grown either in liquid culture medium contained in flasks plugged with cotton wool or on a thin layer of a similar medium solidified with agar agar or less commonly with gelatine.

**Method of Cultivation:** Culture of fungi are usually maintained on conventional agar slants and the agar petridish has been employed in experimental work. Most quantitative studies require a liquid medium which may be either surface (still) or submerged aerated culture. Normal aerobic growth can be obtained only if the inoculum floats on the surface of the liquid. Many spores float naturally, for those that do not use of a bran cultivated inoculum or inclusion of a small amount of agar, gum arabic or poly vinyl alcohol in the medium may be necessary.

Aerial and submerged growth of fungi takes place. Aerial hyphae are remote from the nutrient supply and are bathed in the atmosphere the lower cells by contrast are in close contact with nutrients and with soluble metabolic products but suffer from a partially anaerobic environment. Culture is always deficient in oxygen. This deficiency is shown by the favourable effect on sugar utilization of an increase in surface: volume ratio.

One of the chief advantages, apart from rapid growth is that shaker grown mycelium is usually homogenous enough to be used directly in respirometric and other metabolic studies.

In shake culture each colony is exposed to a uniform environment in all directions, hence the typical colony is a globose structure. Colonial morphology varies however with the species even with in a genus and with the medium. Sporulation is usually another factor of homogeneity shaken culture has a higher content of dissolved oxygen and rate of oxygen diffusion than a still culture for certain purposes, higher rate of oxygen supply may be desireable. Growth on solid media other than agar, is useful in special problems; such material include soil, sawdust, bran and cotton. The use of plant materials sterilized by treatment with propylene oxide is particularly valuable in the induction of sporulation and may have wider appilcation in the study of physiology. Inoculum used for cultures is usually a population of spores from an agar slant. A common occurrence in the fungi is the replacement of the normal type by a variant arising in culture, this development has been analyzed especially with regard to sporulation. Other examples include the loss of biochemical capacities. Change in virulence to a host plant and change in assimilatory capacity. Methods of obtaining single spore cultures of fungi are reviewed by Hilderband (1938). Such cultures may be necessary in order to recover or purify the desired type. Some spores are multinucleate and may contain genetically different nuclei.

Three physiological aspects of the inoculation process may be mentioned. First growth is more rapid with a heavy inoculum than with a light inoculum, in addition, a heavy inoculation of a submerged culture usually results in the development of small colonies, with less danger that oxygen diffusion into the colony will limit growth or respiration.

The use of old spores of low viability is to be avoided, both the age and the conditions of growth of the inoculum, especially medium and temperature. Finally a large inoculum may be required in order to initiate growth on an unfavourable medium.

Mechanically macerated mycelium can be used as inoculum. This is often the method of choice for fungi which do not sporulate. Continuous flow apparatus may have some general interest.

During growth the fungus absorbs food material from the medium and various metabolic products pass out from the hyphae so that the inital composition of the medium changes as the colony develops, and the final composition of the medium may be different from that at the beginning of the experiment. This fact however may be turned to advantage, and much useful information has been accumulated by the analytical observations of the changes taking place in the medium.

**Water availabilty:** Living organisms consist largely of water hence if an organism is to grow and increase in volume it has to take up water from the environment. Growing cells and hyphae have walls and plasma membrane that are permeable to water. This means that water can also be lost to the environment, with the possibility that excessive loss of water could lead to desiccation and death. Whether water enters or leaves a cell depends on the difference between the water potential of the cell and that of the surrounding medium. Water moving from a region of high to one of lower water potential.

Water potential is measured in units of pressure usually either in SI metric units as megapascals (Mpa) or in bar. It is the sum of a number of components of which the most important are osmotic potential, matric potential and turgor potential.

Water potential = Osmotic potential + matric potential + turgor potential

The concept of water potential is widely used by plant pathologists and plant and fungal physiologist's and has a semi official status as the correct way of expressing water availability. However many microbiologists concerned with fungi that live in conditions of very low water availability and cause spoilage of foods and other products continue to use an alternative concept that of water activity (aw). This is defined in terms of the ratio between the water vapour pressure of the solution being considered (ps) and that of pure water (pw).

**The Dynamics of Growth:** Growth is increase in either mass of cells or number of cells. Mass dry weight, usually in the filamentous fungi may be deceptive in as much as it can represent in part the accumulation of polysaccharides, lipids or wall materials without any increase in living protolasm. In *Neurospora crassa* protein synthesis appears to continue for a period of time after that increase in weight has ceased Ballentine and Stephens (1951). Typical curves are characterized by three major phases, with transitions between them.

1. A phase of no apparent growth.

2. A phase of rapid and approximately linear growth.

3. A phase of no net growth or of autolysis and decline in dry weight.

Second phase–may represent polysaccharide synthesis only without increase in other cell components, or it may depend on a mobilization of nitrogen from older hyphae and use of it for new growth. First phase that of no apparent growth, has two components; a genuine lag phase before spore germination and a phase in which growth is occurring but is undetectable. Second phase is of rapid growth. The morphological bases of mycelial development is that growth occurs only at hyphal tips, interior cells of the mycelium do not normally contribute to the new growth directly, though they supply nutrients to peripheral cells, especially to the aerial structures in still culture. Some chemical events also takes place during rapid growth in utilization of carbohydrate, nitrogen and phosphate. Metabolic products i.e. acids may or may not appear in the medium at this time, respiratory activity is at a maximum. Third and last phase is characterized usually by a decline in mycelial weight and in the appearance of nitrogen and phosphate in the medium. Loss in weight varies from negligible to extreme a common pattern is a rapid loss of weight for a short time with no further change thereafter Jermyn, (1953) Autolyzing mycelium of *Penicillium griseofulvum* undergoes an extensive breakdown of chitin. Carbohydrates and protein catalyzed by enzymes of the fungus. Other autolytic products include ammonia, free amino acids, organic phosphorus compounds and sulfur compounds. Cells at the end of the growth phase appear vacuolate, younger cells have a denser and more homogenous protoplasm. Such old cells of course are found even during the growth period in regions remote from the growing hyphal tips. Cessation of active growth appears to be determined by either of two factors. In concentrated media, toxic metabolites can be shown to accumulate and these materials earlier called 'staling substances' demonstrably retard growth. Probably organic acids in high carbohydrate media and ammonia in high nitrogen media, are the compounds most frequently involved.

**Dimorphism:** Several fungi pathogenic to man are found in infected tissues in a unicellular yeast like form, but when cultivated at room temp, grow out in a mycelial form. The term dimorphism strictly applies to these forms only but related phenomena in the normally filamentous fungi may provisionally be considered in the same category. A dimorphic fungi is one in which a reversible tansformation from a mycelial to nonmycelial and unicellular growth type occurs. Dimorphism is common in yeasts and unicellular bacteria may be caused experimentally to form filaments. Some normally filamentous fungi, e.g. Smut fungi grow in culture as unicellular budding cells.

The yeast phase of *Blastomyces dermatetidis* and *Paracoccidioides brasiliensis* is multinucleate that of some other dimorphic fungi multinucleate.

**Temperature and Growth:** Temperature affects growth, spore germination, reproduction and indeed all activities of the organism. Most fungi make at least some growth over a 25 or 30 degree range of temperatures, but narrower ranges are known e.g. in *Merulius lacrymans*. Narrow ranges are often associated with low optima, Many fungi are probably world wide in distribution i.e. their growth is limited by factors other than temperature. Species of *Allomyces* on the other hand are clearly confined naturally to the tropical and warmer temperature regions of the earth. The few marine fungi studied so far have temperature optima for growth that are some what higher than the temperature of the sea. The occurrence and dominance of fungi in decomposing plant materials are largely determined by temperature.

Dermatophytic fungi grow best in culture at 25-30°C i.e. below the host body temperature, fungi causing mycoses grow well at 37°C. It is significant that *Aspergillus fumigatus* a thermotolerant species can cause disease in warm blooded animals.

Surprisingly many fungi are able to grow at 0° or slightly less. Plants under snow may be infected by the parasitic snow molds *Typhula* sp., *Fusarium nivale* and unidentified basidiomycete. *Phacidium infestans* cause of a disease of pines is able to grow in culture at 3°C although its optimum temperature is about 15°C.

A second ecological group of fungi tolerant of low temperature is commonly associated with the spoilage of refrigerated foods. Fungi of many different genera fall into this group and some especially strains of *Cladosporium* and *Sporotrichum* grow at temperatures below zero –5 to –8°C.

Most fungi are unable to grow at 35—40°C. One group includes saprophytes-especially Coprophillous forms and several wood destroying fungi. The second group true thermophiles is defined by the ability to grow at 50°C higher and the inability to grow at temperatures below about 30°C. Truly thermophilic forms also are found among the actinomycetes. The obligately thermophilic *Micromonospora vulgaris* is distinguished by the high resistance of its conidia to heat damage.

**pH and Growth:** Under given conditions, a fungus will grow maximally over a certain range of initial pH values of the medium, and will fail to grow at high and low extremes. Acidity may also affects the entry of essential vitamins, surface metabolic reactions, entry into the cell of organic acids or the uptake of minerals. In contrast the bacteria and the actinomycetes, fungi are relatively more able to invade acid environments. In culture, the larger basidiomycetes are often unable to grow at an initial pH about 7.0 examples include most species of *Marasmius* and *Tricholoma*.

In most physiological studies it is necessary that pH can be controlled atleast within limits. Periodic addition of alkali and the use of soild calcium carbonate in the medium are satisfactory for certain problems. The internal pH of the fungus cell and the fungus cell is usually large enough so that the concept of pH may legitimately be applied.

**Oxygen and Growth:** Fungi are commonly thought of as strictly aerobic. Most fungi grow as well as 20-40 mm oxygen pressure as at atmospheric (160 mm) Denny, (1933); Golding, (1940). However the dry weight is usually affected at somewhat higher oxygen pressure, growth of *Ophiobolus graminis* is reduced at 105mm oxygen. *Aspergillus oryzae* grows best at oxygen pressure higher than atmospheric Tamiya, (1929) Low oxygen requirement for growth fungi successfully colonize environments in which oxygen is limited e.g. relatively stagnant aquatic milieu. The number of fungi usually falls off at lower soil depths Brierly *et al.*, 1928; Burges, 1939 but this may be caused more by high $CO_2$ than by low oxygen. The adverse effect of flooding the soil on the growth and survival of fungi is however believed to reflect a deficit in oxygen.

**Carbondioxide and Growth:** High $CO_2$ presssures generally inhibit the growth of fungi but the level at which inhibition appears in quite variable. Thus *Alternaria solani* is markedly inhibited by 38mm of $CO_2$ and of 150mm or more Bavendamm, 1928; Golding, 1940.

**Water and Growth:** Fungi require relatively high moisture levels, but most of the higher fungi can grow in the absence of liquid water. Relative humidity effects must of course be determined by growing the organism on a substrate. Agar and other solid substrates bran, wood, blocks, fabrics etc. Probably the amount of water in the substrate is the more important, if cotton and wool or incubated at 92 per cent relative humidity only the wool, which absorbs more water at this humidity than does cotton, supports the growth of fungi. Tolerance to low relative humidity appears, as expected to be associated with tolerance to high osmotic pressure Hayashi, 1954 Kaess and Schwartz, 1935 sodium chloride is inhibitory to some fungi Moyer and Coghill (1946) Tolerance of high osmotic pressure is specific. Thus, a mutant of *Neurospora crassa* is inhibited at 0.17 M glucose, at the other extreme species of *Aspergillus* can grow in media with an osmotic pressure of as high as 100 atmospheres. Most fungi cease growth or are markedly inhibited when the concentration of soluble sugar exceeds about 20M. Obligate halophily an absolute salt requirement has been reported for several fungi Malevica, 1936; Vishniac, (1955). Although at least some marine fungi do not require salt.

**Vitamin Requirements:** Fungi like all other organisms require minute amounts of specific organic compounds for growth. Most of the known vitamins have a catalytic function in the cell as coenzymes or constituents parts of coenzymes. For the essentiality of vitamins in those fungi which do not require any external supply is found in the fact that deficient mutants can usually be obtained by appropriate techniques, in addtion to the classical *Neurospora crassa*, include speices of *Penicillium* and *Aspergillus* (Bennett, 1921, Gaumann and Nef (1947); Meyor (1955), genera in which naturally occurring vitamin deficiencies are uncommon.

Environmental factors conditioning vitamin deficiencies include temperature, acidity and salt concentration. A mutant of *Neurospora crassa* for example requires riboflavin only at relatively high temperature. *Sclerotinia camelliae* requries more inosital at 27° Barghoorn and Linder, (1944). *Sordaria fimicola* requires thiamine only if the initial pH of the culture is less than 4.0, possibly because thiamine synthesis is inhibited by hydrogen ion Horowitz, (1951) *Pythium butleri* requires exogenous thiamine only in a high salt medium.

Several fungi *Alternaria solani* Hosten et al., (1953), *Fusarium solani* Stockdale, (1953) and *Penicillium digitatum* Crasemann, (1954) are reported to be accelerated in early growth by vitamins the requirements for which are not evident if total growth over a long period is measured. In *Myrothecium verrucaria* a requirement for biotin is detectable only at a stage shortly after spore germination Kaess and Schwartz, (1935). Among the dimorphic fungi, a single study suggests that the yeast and the mycelial phases of some forms differ in their biotin requirements. Multiple requirements are common in yeasts five or six vitamins may be needed. Among the filamentous fungi, there are several instances of a requirement for three different vitamins, e.g. in *Ascoidea rubescens* Day and *Harvey*, (1946), *Blastocladia pringsheimi* Brooks and Books, (1941) and *Trichophyton discoides* strains (Jilson and Nickerson, 1948).

**Composition of Fungs Cell:** The theoretical significance of chemical composition is some what limited. The qualities of fat, carbohydrate, ash, wall material, and total nitrogen are all more or less responsive to the culture medium. The Principal difference between fungi and bacteria is in the high chitin content of most fungi chitin is present in the bacteria, green algae do not have chitin walls. Lipid accumulation is more common and attains higher levels in the fungi than in most bacteria.

**Major constituents of the wall:** Growing fungus cell indicate that 85-90 per cent of the fresh weight of the mycelium and fleshy sporophores in water. Leathery sporophores and sclerotia have less water. The conidia of most fungi appear to have a relatively low water content 17.4 percent in *Aspergillus oryzae*, 25 per cent of *Monilia fruticola* and 6 per cent in *Penicilluim digitatum*. Most of the water in the spores is hygroscopic.

Elementary composition of fungi has been studied chiefly with respect to carbon and nitrogen. The carbon content of a number of fungi was 40-44 per cent of the dry weight. In *A. oryzae* it is about 49 per cent, *A. niger* 46 per cent. Most fungi are 40-50 percent carbon. Major constituent of fungi and actinomycetes are Carbohydrate, Chitin, Protein, Lipid and Ash. Carbohydrate may be from 9-60%, Chitin-5.5-10.6 percent, Protein 12.0-38.8 per cent, Lipid 2.20-22.4 per cent and Ash 5.35-6.24 per cent.

**Mineral Constituents of Cell:** Ash content of fungi varies with the medium used and presuambly with species. In *Rhizopus japonicus* ash content is 5.52. *A. niger* 2.39 and *A. oryzae* 7.19, *Penicillium notatum* 28.3-34.0 per cent. Fruiting bodies collected in the field may be as low as 1.08 or as high as 29.8 per cent ash. most have 5-10 per cent. The total ash content and the concentration of individual elements are influenced by the environment and by the stage of development. The total ash and the concentration of individual elements are also influenced indirectly through effects on growth or pH, by the type of nitrogen source and by the concentration of nitrogen. A low nitrogen supply may be expected to act directly on phosphorus content by restriction of the synthesis of nucleic acids. Only phosphorus, sulfur and chlorine occur in fungi in organic combination. Phosphorus is present in several types of organic molecules--sugar, phosphates, phosphoproteins, nucleic acids and phospholipids.

The inorganic phosphate of fungal mycelium include orthophosphate and inorganic condensed phosphates. Pyro-phosphate occurs in *A. niger*, as well as metaphosphate is also found in *A. niger* and *Neurospora crassa*.

**Cell Wall:** Morphologically, wall may be a multiple structure with layers of different composition. In most fungi the basic material of the cell wall is chitin, a polymer of N-acetyl glucosamine, identical chemically with the chitin of arthropods chitin is found in the zygomycetes and in the higher fungi ascomycetes and basidiomycetes, representative of myxomycetes, oomycetes and monoblepharidales were found not to form chitin.

In actinomycetes chitin is absent, the sheath surrounding the stalk of *Dictyostelium discoideum* is composed of cellulose, laid down extracellularly. Callose appears to be an acidic polyglucose, its relation to the cell wall itself is uncertain and cannot at present be accepted. Lignin occurs in relatively large amounts in several fungi the identification rests on solubility characterstics alone and it seems very doubtful that the substance is chemically the same as the lignin of higher plants.

**Cellular Carbohydrates:** Polysaccharides and related compounds of chitin are associated with the cell wall and other appear in the medium as well as the mycelium. Cells grown on high carbohydrate media are rich in polysaccharides. The first fungal polysaccharide was obtained from *Penicillium glaucum* spores named spore starch. The most common sugar component is glucose, the next common galactose. The glycogen of fungi is chemically identical with or very similar to animals.

Polysaccharides of which the constitution is not fully known include especially the hemicelluloses, defined by their solubility and ease of hydrolysis. The crude hemicellulose fraction of *A. niger* constitutents 12 to 31 per cent of the dry matter of the mycelium, and is like other constituents of the mycelium, affected by the supply of the metals in the medium. Pentosans have been isolated as about 1 per cent of the dry matter of fungi. Mannans are present in yeast but not common in fungi.

**Nitrogenous Constituents:** The total amount of nitrogen in mycelium is usually determined by kjeldhal method. Total nitrogen of cultivated fungi range from 2.27 per cent of the dry weight, in *Coprinus radians* to 5.13 per cent in *Trichoderma lignorum* sporophores from nature had as little as 1.56 per cent nitrogen. Young mycelium is higher in nitrogen than old mycelium during autolysis both proteins and other nitrogen compounds are liberated into the medium. The nitrogen source used also exerts an influence on the total nitrogen of the mycelium in general, the nitrogen content is higher when ammonium salts are used than when nitrates provide the nitrogen. Qualitatively the amino acids of fungal poteins are much the same as of other proteins.

**Cell lipids:** Total lipid content of fungus mycelium usually expressed as a percentage of the dry weight, is strongly affected by cultural conditions and age. High fat values 20 per cent or more of the dry weight are found among species and strains of *Aspergillus, Endomyces, Fusarium, Mucor, Penicillium* and *Torula* and in individual species of other genera. Conidia of common saprophytic fungi do not have usually high fat contents, but the sclerotia of *Claviceps purpurea* contain 43 per cent lipids and uredospores of *Puccinia graminis tritici* are high in fat.

**LIPIDS OF FUNGI**

| Organism | Iodine Number | Acid Number | Phosphatides | Unsaponifiable Matter | Sterols |
|---|---|---|---|---|---|
| *Aspergillus niger* | 95.1 | 71.2 | – | 12.0 | 1.4 |
| *A. sydowii* | 114.4 | 43.4 | Present | 8.2 | 5.4 |
| *Blastomyces dermatitidis* | 106.1 | 45.3 | 24.3 | 8.0 | – |
| *Fusarium graminearum* | 84.7 | – | – | 2.1 | – |
| *Penicillium javanicum* | 84.0 | 10.6 | –2.0 | – | |
| *Phycomyces blackesleeanus* | – | – | 9.9 | 5.3 | 4.0 |

Content of free acid is markedely affected by the cultural conditions. In *Penicillium javanicum* the acid number rises from 10.6 to 50.7 as the glucose level of the medium is increased from

20-40 per cent. Any factor which increases fat formation also increases the relative amount of free fatty acids. *Aspergillus nidulans* lipids are very low in free acid, as are sclerotia of *Claviceps purpurea*. Lipase action during preparation of the mycelium may contribute to the high apparent acid value of fungal lipids. A number of fatty acids which are not known to be constituents of fats have been isolated from fungi. Three of these agaric, spiculisporic and mineoluteic acids are long chain hydroxy-tricarboxy acids, similar to some of the lichen acids. Phosphatides determined by their solubility are usually less than 10 per cent. The phospholipid fraction of *Penicillium chrysogenum* yields inositol phosphate and an inositol phospholipid occurs in *Neurospora crassa*.

The mycelium of *Asperillus sydowii* contains at least 0.1-0.4 percent of cerebrin this partially characterized nitrogenous lipid is probably identical with that isolated earlier from yeast, *Claviceps purpurea* and from the sporophores of several basidiomycetes. Virtually all fungi appear to form sterols. The amount in mycelium is usually about 1 per cent less of the dry weight. The principal sterol of fungi is ergosterol–a dihydroxy derivatives, fungisterol occurs in ergot. Ergosterol occurs as the palmitate in *Penicillium* species and *Aspergillus fumigatus*.

# 5. REPRODUCTION

Fungi produce enormous number of spores in beautiful and elegantly adapted strcutres, by means of sexual or asexual means in nature. We observe such characters by simulating natural environment under laboratory conditions, which fascinate us to study them. Morphologically, sexual and asexual stages produces by a species are often very different. Sexual stage is more complex as compared to asexual stage and requires utilization of more resources than simpler conidiophore forming asexual stages. There are many fungi in which only asexual reproduction is known e.g. Fungi imperfecti (**Deuteromycotina**) and there are many which reproduce only sexually.

In sexual reproduction, nuclear fusion and meiosis are essential features, while asexual reproduction involves only mitotic divisons and serves to multiply genotype. In sexual reproduction, genetically the fusing nuclei differ sexual meiotic products can be recombinants. Sexual reproduction generates variations and new combinations of characters can arise in populations.

During reproduction, morphological vegetative growth phase is transformed into physiologically specialized reproductive unit i.e. spore. These spoes are actually vegetative cells except having multilayered thick walled, sometimes impregnated with melanin like pigments and lipids. Spores cytoplasm, endolpasmic reticulum and mitochondria are poorly developed and differentiated. Spores also contain substantial amount of storage materials like lipids, glycogen, trehalose, low water content and low rate of endogenous metabolism. The manner of spore development and morphological characters of the fruiting bodies in case of higher filamentous fungi is used as an important character for identification of fungi. The formation of reproductive organs where entire thallus (fungal cell) is converted into one reproductive structure are called **Holocarpic** (chytrids), while in the majority of fungi reproductive structures arise from a portion of the thallus while remaining tissue continues its normal somatic or vegetative activities. These fungi are called as **Eucarpic**.

Life cycle of a fungus involves plasmogamy followed by karyogamy and meiosis which occur at regular intervals and it is completed only when sexual fusion occur. The cycle lacks in asexual or vegetative reproduction. The perfect state in fungi is characterised by sexual spores (ascospores, basidiospores etc) while imperfect state is one which is characteised by asexual spores or even absence of sproes. Hennebert and Weresub (1977) consider holomorph of a fungus as one which includes all forms and phases, while telomorph for sexual form (e.g. characterized by ascocarps and basidiocarps) and anamorphs for asexual form (e.g. characterized by conidiomorph).

Synanomorph is applied to any one of two or more anamorphs which have the same telomorphs (Gams, 1984) e.g. in case of black stem rust of wheat. Reproduction in fungi is basically categorized into two types-asexual and sexual. While the vegetative reproduction is a mode of asexual one, which is considered here separately.

## Vegetative Reproduction

This is one of which a specialized vegetative cell or group of cells or even aggregate mass of vegetative cells help in perpetuating vegetative structure of the fungus. This includes:

A. Fragmentation; B. Fission; C. Budding; D. Arthrospores; E. Chlamydospores; F. Gemmae; G. Sclerotia; H. Rhizomorph; I. Mycelial cords; J. Pseudorrhiza; K. Pseudosclerotia.

**Fragmentation:** The fungal hypae break into small pieces (not as individual cells) and each piece may later grow into a new mycelium, e.g. common in soil fungi and also in growth of culture from marginal leader hypae.

**Fission:** A common method of reproduction in yeasts. The yeast cell divides into two daughter cells which separate by constriction or transverse wall.

**Budding:** Budding is production of small outgrowths (buds) from the present cells. During bud formation, nucleus from the parent cell divides mitotically, and daughter nucleus migrates into the bud. Bud increase in size while still attached to the parent cell and eventually breaks off (due to disturbance in the environment or physiological changes during it's metabolism) and forms new individual. Separation of the bud from the mother then takes place and bud scar is evident on the mother cell, Opposite a birth scar on the daugther cell. A birth scar resembles a crater on the surface of the mother cell with a raised circular rim partially consisting of chitin. Chains of buds forming a short mycelium (pseudomycelium) are produced in nutritive rich medium.

**Spores Arthrospores:** (arthro-jointed, formed in chains by the simultaneous or random fragmentation of the hypha). This is a modified type of fragmentation in which hypha ceseases it's apical growth. Then the septa are laid down in the tip ring. these septa are two-ply which makes separation of the two layers leading to the disarticulation of the terminal part of the hyphae into cylindrical segments referred as **Arthrospores** or Arthroconidia (Cole and Kendrick 1968, Kendrick 1973) Since there is no specialised cell producing these arthrospores, legitimately now it is refered as arthrospore rather than Arthroconidia.

**Spores Chalmydospores:** (Gr.- Chlamy - mantle; spora - seed, spore) Chalmydospores originate by the modification of the hyphal segment and possesing an inner secondary wall usually impregnated by hydrophobic material (Griffith 1901). Under unfavourable conditions the terminal or intercalary segment of the vegetative mycelium becomes packed with food reserves and develops thick walls. The walls may be colourless or pigmented with dark melanin pigments. Thus spores become separated from each other by the disintegration of intervening hypae. This type is sometimes referred as thick walled thallic condition that generally functions as **resting**

**spore**. These spores survive under adverse environmental conditons such as high and low temperature, lack of moisture, exhaustion of food supply and so on. Since smut teliospores are formed somewhat in a same manner of chalmydospores, some authors actually refer to these (smut spores) as **chlamydospores**.

**Gemmae (L - gemma - bud):** Gemmae are specialised chalmydospores which are borne terminally either singly or in chains which are separated after maturing by breaking free from the mycelium and are dispersed in the water currents in case of the aquatic fungus like *Saprolegnia*.

**Sclerotia:** Pseudoparenchymatous aggregation of the vegetative hyphae which serve the function of survival is referred as **sclerotia**. These are common in plant pathogenes like *Thanatephorus cucumeris* and species of *Claviceps; Balansia* and *Sclerotinia*. The form of sclerotia is variable. The sclerotium of *Polyporus mylittae* (Australian native black fellows bread) can reach the size of a man's head and is considered as edible. On the other hand, some sclerotia may be of microscopic dimensions consisting of a few cells only.

Development of sclerotia is categorised into several types such as loose, terminal, strand, globose or complex multicellular types.

**The loose type** (*Rhizoctonia solani*)**:** Sclerotia are thin brown scurty scales on the surface of the potato tuber. Cell of the sclerotium is barrel shaped and considerably wider than vegetative hyphae. These cells become filled with dense contents, numerous vacoules, darkened to reddish brown without internal differentiation of zones of tissue.

**The terminal type** (*Botrytis cinerea and B. allii*)**:** Sclerotium found on the overwintering stems of infected umbelliferous plants (*Anthriscus, Haracleum* and *Angelica*). Sclerotia arise due to repeated dichotomous branching of the hypae accompanied by cross wall formation. Hyphae then coalesced to give the appearance of a soild tissue. Mature sclerotium is 3-5 mm, usually flattened, measuring 1-3 mm in thickness. Sclerotia are present parallel to the long axis of the host plants. Outer ring composed of several layers of rounded, dark, thickened cells and narrow cortex of thin walled pseudoparenchymatous cells with dense contents and medulla consists of loosely arranged filaments.

**The strand type** (*Sclerotinia gladioli and Sclerotium rolfsii*): Scletotia are 0.1 to 0.3 mm in diameter and differentiated into ring of small thick walled cells and medulla of large thin walled cells. More complex sclerotia are found in *Sclerotium rolfsii* with differentiated four zones, a fairly thick skin or cuticle with remnants of walls attached outside of the empty melanized thick walled rind. A rind made up of 2-4 tangentially flattend cells, a cortex of thin walled cells with dense staining contents and medulla of dense filamentous hypae with dense contents all the cells of sclerotia are with thick walls than those of vegetative cells.

**Other types:** Other types of sclerotia as in case of ergots of grasses and cereals (*Claviceps* spp.) insect-parasitic sclerotia of *Cordyceps militaris*, giant sclerotia of *Polyporus mylittae* or even pseudosclerotial plate or reaction zone lines developed in woody tissue colonized due to wood rotting fungi. Some of these vegetative reproductive structure (sclerotia) may develop into fruit bodies such as apothecia emerging out of the sclerotia of *Sclerotinia spp.* or perithecial stroma in overwintered sclerotia of *Claviceps purpurea* or basidiocarp developed from buried sclerotia of *Polyporous mylittae.*

**Rhizomorphs** (*Armillariella mellea*): The formation of aggregates of parallel, highly differentiated hyphae with well developed apical meristem, a central core of large thin walled elongated cells and a ring of smaller, thicker walled cells with dark melanin pigmentation. The fungus spread from one root systems to another by means of the rhizomorphs. There are two types of rhizomorphs found, one dark cylindrical type measuring up to 4 mm in diameter and other paler, flatter type common beneath the dark of infected trees. The terminal half milimeter of the rhizomorph contains compact growing point if isodiametric cells, protected by a cap of interwined hyphae is slimy matrix. Mitotic activity is being reported in this meristematic zone after behind of which is a zone of elongation. The middle part of the mature rhizomorph is hollow and filled with air or even may be solid. Surrounding the central lumen is a central medulla having zone of enlarged, elongated hyphae, four to five times in diameter than the vegetative hyphae which serve the tanslocation of nutrients. Towards the periphery of the rhizomorph the cells become smaller, darker and thick walled, extending outwards between the outer layers of cells of the rhizomorphs. Growth of vegetative hyphae resembles the root hair zone of that of the higher plants. Ultrastructure of hyphal structure of rhizomorph tip is meristem like and would be totally alien to the growth strategy of the fungal hyphae as the central apical initial produces axially arranged tissues This tissue is sometimes referred as Meristemoids.

**Preservation:** Peservation of fungi is to maintain a viable inoculum in a non growing condition. Culture and herbarium specimens often survive for surprisingly long periods after drying, by the formation of long lived spores. Zoble 1943 records survival of spores upto 21 years, he found that survival of dried cultures is common in ascomycetes and basidiomycetes, but the Mucorales are relatively short lived. The sclerotia of several fungi, which may lived for as long as 13 years are less viable if stored dry. A few of the more leathery basidiomycetes sporophores survive dry for a period of years; more succulent types are less durable.

In the soil tube method a fungus is transferred to moist sterile soil, pH 6-7 with calcium carbonate and grows and sporulates until the soil dries out. Some fungi survives at least 5 years in soil tubes, too many do not live that long and the method therefore is to be regarded as a special one not adapted for general use.

In mineral oil preservation sporulation is not essential, cell components, or it may depend on a mobilization of nitrogen from older hyphae and use of it for new growth. The morphological basis of mycelial development is that growth occurs only at hyphal tips. Interior cells of the mycelium do not normally contribute to the new growth directly, although they supply nutrients to peripheral cells, especially to aerial strucures in still culture. Growth in any one time period is therefore a function, not of the total yeasts but of the number of hyphal tips and of the rate at which these tips are supplied nutrients.

Lyophilization of spores suspended in serum, gelatin, dextran or other colloidal materials appears to be a usable method, although again not successful with some fungi e.g. Entomophthorales and the Dermatophytes. Simple vaccum drying without freezing is suitable for many fungi. For sporulating fungi, lyophilization appears to be the best single method.

**Mycelial cords:** Filamentous aggregation of relatively undiffernetially parallel hyphae develops due to concentrated growth of individual hyphae gradually building up a film mycelial framework. Mycelial cord vary in complexity from relatively loose association of morphologically similar hyphae to well defined anatomically complex structure with a marked degree of internal differentiation. Thin walled as well as several other hyphal types are also present. Calcium oxalate crystal which is excretory as well as antibiotic in function is heavily encrusted outside the hyphal walls. Internally coiled thin walled tendril hypha bind structure together with thick walled fibrous hyphae. Thus, fibrous hyphae hold tensile strength of the cord, large orange colour vessel hyphae with incomplete septa help in translocation. Anastomosis between the hyphae of the major strands helps to consolidate them and they may further increase in thickness by the accretion of minor cords. Mycelial cords are capable of translocating materials in both direction. Fungus extends from the established food base and colonizes new strata. These mycelial vegetative aggregates thus increase the inoculum potential of the fungus at the point of the colonization.

Mycelial cords and rhizomorph are extreme in a range of linear aggregation doing the function of vegetative reproduction.

**Pseudorhiza:** The mycelium of some mushrooms (*Cedemansiella radicata*, species of *Termitomyces*) originate from the materials buried at a considerable distance underground and in becoming elevated to the surface to form the fruit bodies (mushrooms). It becomes linearly aggregated into one or more stout cylindrical colours each of which bears a fruit body at the ground level. Pseudorrhiza is a root like continuation of the stipe of the mushroom connected to specific underground source of nutrients.

e.g. *Oememansiella radicata* - root of higher plants; *Collybia fusipes* - buried wood or roots of higher plants; and all species of *Termitomyces* - live termite nest.

**Pseudosclerotia:** Sclerotium like mass of friable materials of the substratum bound together in a hard mass of mycelium. Australian stone making fungus *Polyporus basilapiloides* having 20 x 10 cms. While smaller pseudosclerotia are common in mushrooms, morels and Gasteromycetes growing in the sand. In parasitic species of fungi, these pseudosclerotia consist of mycelium plus remnants of plant or animal tissue of the host e.g. species of *Cordyceps* parasitizing caterpillers and *Monilinia/Sclerotinia* parasitizing stone fruits and forming hardened mummified mycelium. Such "mummies" or pseudosclerotia gives rise to fruitification after overwintering.

## Asexual Reproduction

It is defined as the non sexual production of the specialised reproductive cells, Spores, Or it is a method of propagation which do not involve fusion of compatible nuceli followed by karyogamy and meiotic division. Some authors also include production of somatic cells or group of cells by simple fragmentation, fission, budding or perennating individuals (which are discussed here under vegetative reproduction) Only a method of propagation with delimit a specialized reproductive cell (Spore), that is capable of germinating into germ tube that grew into independent mycelium is referred legitimately under asexual reproduction. In this reproduction numerous identical individuals are produced which takes place several times during favourable season. Infinite variety of spores makes fungi particularly fasinating. Spores are thus typical features of fungi that serve for multiplication, dispersal or survival. The manner of spore development and morphological characters is used as an important criteria for identification of fungi. Based on ornamentation these are referred as reticulate, undulate, echinulate, warty, verrucose, alveolate, punctate etc. L.O. analysis

(Lux/light obscuritas analysis/darkness), a method introduced by Erdtman (1943) for pollen is also used for fungal spores (Payak, 1959). By this mehtod different types of surface ornamentation are distinguished microscopically due to difference in the apperance of the spore surface at the upper and lower focus. Based on the septation in spores one can use the terms like Amerospores (one celled), Didymospores (two celled), Phragmospores (two or more transverse septa), Dictyospores (transverse and longitudinal septa), Scolecospores (vermiform or filiform), Staurospores (radiate or stellate) and Helicospore (spirally coiled).

The shape of the spore is also important in character to understand its morphology such as ailantoid, pyriform, obpyriform, fusoid, fulcate, filiform etc.

Even on the basis of ontogeny, spores are classified as exogenous and endogenous spores. Based on the mechanism of spore formation one can use the term **xerospores** - formed in dry masses while **gloiospores** formed in wet shiny viscous droplets or in slimy masses. Asexual spores produced in sacs or in a vessel or in sporangia are referred as sporangiospores. While those which are produced at the tip or the side of the hyphae in various ways are called conidia. Sporangiospores may be motile **zoospores** or **swarm spores** and non motile **alpanospores**. Some more commonly encounted types of asexual spores are:

i. Zoospores; ii. Sporangiospores; iii. Conidiospores

**Zoospores (Gr. zoon - animal; Spore - seed, spore):** The spores which are self propelled by means of flagella. Zoospores are of three kinds:

**Posteriorly uniflagellate:** Zoospores with flagella of whiplash type (Chytridiomycetes).

**Anteriorly uniflagellate:** Zoospores with flagella of the tinsel type (Hypochytridiomycetes).

**Biflagellate zoospores:** Zoospores with anteriorily or laterally attached flagella one of which is of whiplash type and other is of tinsel type (Oomycetes).

The whiplash flagllum is divided into two parts. The basal portion is much longer than the terminal portion which is usually very short and flexible. The tinsel flagellum is a feathery structure consisting of a long rachis with lateral hair like projections termed as **Mastigonemes** or **Flimmers** on all sides along it's entire length. The flagellar apparatus in motile fungal spores is very complex consisting of the flagellum itself, a **Kinetosome** (or **Blepharoplast** or **basal** body) to which the microtubular components or **Axoneme** of the flagellum is attached and **Rhizoplast** (Gr. Rhiza - root plus plastid) or rootlet by which kinetosome is attached to the nucleus of the zoospore by various microtubular elements. In a fungal flagellum the axoneme is of the characteristic 9 + 2 construction. The nine peripheral microtubular components form a cylinder around the two central ones. Each central component is a single tubule, while the nine peripheral components, each consists of two microtubules. A series of microtubules extends from proximal end of the kinetosome towards the nucleus and nuclear cap which provides them structural rigidity. Extending into the mitochondria, linking up the kinetosomes, there are three striated bodies referred as **flagellar rootlets**, or **striated rootlets** or **banded rootlets**. The banded rootlets are contained within separate channels and each is surrounded by a unit membrane. Since energy for propulsion of zoospores is generated within mitochondria, banded rootlets are responsible for transmitting energy to the base of the axoneme. As well as banded rootlets serve to anchor the flagellum within the body of the zoospores.

Sporangiospores may be uni or multinucleate and are unicellular, generally smooth walled, globose to ellipsoid in shape. If spores are dispersed with rain splash or insects, they are surrounded by mucilage and if dispersed by air current, spores are dry (**Xerospores**).

The structure bearing these sporangiospores is important structure for their idetification e.g. sporangiophore may be simple or branched, or whorled or highly branched. Sporangiospores or few spored sporangiola are borne on a central head Sporangioliferous head, which may be single or even branched.

**Conidiospores-conidia:** A specialized non motile asexual spore usually caducous, not developed by cytoplasmic cleavage (sporangiospores) or free cell formation (ascospores), but developed from a conidium initial or a part of cell from which conidium developes. The term conidium is being unfortunately used in a number of different ways,so that it has no longer any precies meaning. There is great variation in conidial ontogeny (**Conidiogenesis**).

## Sexual Reproduction

Sexual reproduction involves union of two compatible nuclei and process typically consists of three distinct phases - plasmogamy, karyogamy and meiosis **plasmogamy** - (Gr. plasma = a molded object; gamous = marriage, union) union of the two protoplasts and brings the nuclei cloes together within the same.

**Karyogamy** - (Gr. karyon = nut, nucleus; gamos = marriage, union) fusion of the two nuclei brought together by plasmogamy is called karyogamy and constitutes the second phase of the sexual reproduction. Sooner or later after nuclear fusion is followed by **meiosis** (meiosis = reduction) which again reduces the number of chromosomes to the haploid and which constitutes the third phase of the sexual reproduction. In more complex fungi, first and second phases are separated in time and space with palsmogamy resulting in a binucleate cell containing one nucleus from each parent. Such pair of nuclei is referred as heterokaryotic diakaryon (NL di = two; Gr. karyon = nut, nucleus) These two nuclei in dikaryon do not fuse for a considerable period in the life history of the fungus. Meanwhile during growth and cell division of the binucleate cell, the dikaryotic condition is perpetuated from cell to cell by simultaneous conjugate division of the two closely assosiated nuclei and by the separation of the two resulting nuclei into the two daughter cells. In a true sexual cycle these three processes occur in a reuglar sequence and usually at specific points. Thus sexual phase in fungi is haploid (gametophytic generation) though dikaryotic is common in basidiomycetes. The methods of sexual reproduction involve the union of two compatible nuclei which may be carried in motile or non motile gametes, in gametangia or even in the somatic cells of the thallus. Plasmogamy is the first phase during sexual reporduction due to sexual anastomosis and may take place through planogametic conjugation, gametangial copulation, gametangial contact, gametangial conjugation, seprmatization, somatogamy or even through spontaneous heterokaryosis.

**Planogametic conjugation:** Planogamete is a motile gamete or sex cell and planogametic conjugation involves fusion of two gametes, one or both of which may be motile. It takes place only when free water is available at this critical stage of the life cycle. Such reproduction is accomplished by one of the following methods:

**Conjugation of the isogamous planogametes** (*Olpidium viciae, Synchyterium endobioticum*): The motile gametes are morphologically similar but physiologically different. Unite in water to form motile zygotes. In some species gametes originate from same gametangia will not fuse.

**Conjugation of anisogamous planogametes:** One planogamete is considerably larger than other. Fusion takes place in water and motile zygote is formed. In this some species of Blastocladiales produces separate male and female gametangia in which even gametes are distinct from each other i.e. male one is bright orange in colour due to a carotene and female one is colourless. But these gametangia are produced on the same branch of the gametophytic thallus thus these are haploid and monoecious. Motile zygote immediately produces into diploid asexual sporophyte resembling sexual gametophyte in general habit.

**Fertilization of a non motile female gamete** (*egg*) **by a motile male gamete** (antherozoids) (Monoblephariales): Male gametes (antherozoids) are released from the male gametangia (antheridia) into the water and swim away. Some of them reach the female gametangia (oogonia) where upon one antherozoid enters each oogonium and units with the egg (oospore, female gamete) within. The flagellum of the spermatozoids (antherozoides) being absorbed within few minutes. Following plasmogamy the zygote (oospore) secrets golden brown membrane around itself and nuclear fusion later occur. Oospore germinate under favourable conditions by producing a hyphae that developes into a new thallus.

**Gametangial copulation** (*Rhizophydium couchii, Chytridium sexuale*)**:** The process of sexual reproduction involves the transfer of entire protoplast of one gametangium into another one. In this members female thalli appears to be nothing more than a sporangial thallus with its exogenous growth arrested at an early stage by the enlargement of the male thallus. Later encysted male cell empties its contents into female thallus and the combined protoplasts moves into an endobiotic swelling increasing in size and soon becoming surrounded by a thick warty wall (oospore).

**Gametangial contact** (Oomycetes, some ascomycetes, *Pyronema*)**:** This type of union is heterogamous, but neither male nor female gamete is motile i.e. reproduction is always heterogamete. The male and female gametangia come into close contact and the male gamete (non motile) consisting mainly of the nuclear material are directly transformed into the female gametangia through pore dissolved in a common wall at the point of contact - or even through an short fertilization tube (trichogyne). Oogamous condition which results into the formation of the non motile zygote (oospore) is a specialised form of gametangial contact because here the male gamete (nuclear material) fertilizes one or more oospore (egg within the oogonia). The oospore may be centric, ecentric, subcentric due to presence of fatty reserves in the form of oil droplets outside the ooplast. While in higher fungi (Ascomycetes) there are no oospores and female gametes are represented by nuclei. After fertilization these female nuclei associated in conjugate pairs with the male nuclei and dikaryotic tissue (properithecium or propseudothecium) are formed and nuclear fusion (karyogamy) eventually takes place in the ascus mother cell to form true zygote and diploid nuclei. The terms used in Ascomycotina are antheridia and ascogonia for respective male and female gametangia. Since both gametes are non motile, gametangial contact is possible in non-aquatic fungi and fungus achieves a measure of independence from free water in situation which are liable to dry up.

**Gametangial conjugation** (Fusion of similar cells for reproduction) **Zygomycotina:** Again there fusion of two multinucleate gametangia that are mainly similar in structure but may differ in size. The first step leading to the formation and eventual fusion of these gametangia involves the formation of specialized hyphae (Zygophore). These structures are produced in actively growing vegetative hyphae and it is chemically induced process. Compatible zygophores are attracted one to the ohter and fuse in pairs at their tips to form a so called fusion septum. The tips of the two zygophores swells to form progametangia (Gr. pro-before + gametangium) Another septa are formed near the tip of the progametangia separating the two cells, terminal gametangium and a suspensor cell. The fusion septum then dissolves, the protoplast of two gametangia mix, and karyogamy externally takes place. The cell formed by the fusion of the two gametangia is initially referred as the prozygosporangium. It enlarges, develops a thick multilayered wall and becomes the zygosporangium in which a single zygospore develops.

Zygospore is a typically large thick walled warty structure with large food reserves and are unsuitable for long distance dispersal. (**Xenospores**) They usually remain in the position in which they are formed, awaiting suitable conditions for further development. In zygospore nuclear fusion and meiosis eventually occur, may have one or two or many nuclei within it. Nuclear fusion may occur early or may be delayed until shortly before zygospore germination.

**Spermatization:** Spermatization consists of copulation of spermatia or microconidia or oidia produced in special receptacles (spermagonia), with the female sex cells. The ascogonia are prouduced in the ascocarp initials, through the medium of the receptive hyphae (trichogyne) through which the spermatial nucleus reaches the ascogonia where it pairs with the female nucleus resulting in plasmogamy. The spermatia are minute, bacillar or helicoid, uninucleate cells incapable of germination. The sperm cells are produced in mucilage and with absorbtion of moisture, escape through ostiole and are carried to the compatible trichogyne through the external agencies like insects and rain drops. This is the most conventional pattern of spermatization reported in many species of Ascomycetes, known as spermatia trichogyne copulation. *e.g.* species of *Mycospharella, Gnomonia, Chochliobolus, Neurospora, Sclerotinia*, members of Loboulbeniales, Lecanorales, *Elsinoe thirumalacharii, Rosenscheldiella eugeniae, Phyllachora symplocicola, Microcystis indicus*. [Kamat and Pande (Chiplunkar) 1972]

Monilioid conidia and oidia (microconidia) are also reported to function as male cells that bring about fertilization behaving like spermatia. This pattern is known as conidization or oidization in *Neurospora sitophila*. However in this species microconidia play dual role of sex cell (spermatia) as well as producing normal mycelium.

In case of *Phyllachora actionodaplnae* and *Cyclotheca kamatii*, spermagonia which are round or flattened structure, lie completely embedded in the host tissue, but in close proximity with ascogonial chamber which develops at approximately the same level as spermagonia, within the host leaf. These ascogonial locules develop several specialised, enlarged, deeply staining nature cells within the cavity along the basal layers. The spermatia from the centrally situated spermagonium are released into the ascongonial initial, through the basal canal that communicates the spermagonium with ascocarp initials on either sides. The spermatia so relased diplodize the specialised ascogonial cells which are intially monokaryotic but later become dikaryotic in status. The process of spermatization involves self fertilization since both the sex organs are produced on the same thallus. (monoecious).

The dikaryotic phase resulted after ferilization may persist for same time or even may extend to other cells to form dikaryotic tissue, but ultimately leads to development of ascus or basidium (teliospore) are karyogamy. Process of spermatization is known only in Rust fungi (uredinales) in which spermatia and the trichoygne (receptive hyphae) are produced from the same spermagonia (receptive hyphae) are prouduced from upper wall layers of spermagonia. The spermatia and receptive hyphae produced from same spermagonia are usually not compatible since most of the rusts are heterothallic. And after spermatization this dikaryotic hyphae initiates aecia.

**Somatogamy:** Anastomosis of somatic hyphae is a very common phenomenon in higher filamentous fungi and it has a sexual significance only when it brings together compatible nuclei of opposite mating type into one cell. This results in dikaryon formation (plasmogamy) and this dikaryotic cell produces dikaryotic hyphae, secondary mycelium and ultimately differentiate into dikaryotic tissue producing fruiting body in which karyogamy occur to form diploid nuceli. This type of reproduction is seen in terrestrial fungi which do not require free water for fusion. This type of reproduction is regarded as reduced but highly efficient form of sexuality.

In homobasidiomycetous fungi hyphal fusion of two compatible hyphae is controlled by chemical signalling mechanism referred as **Horming mechanism**. During this process curvature of hyphae takes place prior to hyphal fusion. The secondary mycelium developes after fusion continues to grow through a special structure referred as **clamp connections** that are formed during nuclear division. The presence of clamp connections is generally indicative of the dikaryotic condition. There are some exceptions, where, since homokaryotic mycelia of some species are known to have clamps while dikaryotic mycelia of other species apparently lack them (Raper 1966).

Somatogamous fusion of compatible vegetative cell which is always under hormonal control and is common phenomenon is yeast fungi which results in the diploid condition or diploid yeasts of even directly into ascus. In some yeasts adjacent ascospore of the opposite mating type of the same ascus fuses followed by plasmogamy and karyogamy within ascus by a tube which push through ascal wall and will act as sprout mycelium (diploid) from which yeast cells are budded. These buds are seprated by septa and soon become sprout cells. Meiosis converts these sprout cells (somatic haploid yeast cells) into ascus each producing ascospores with two of each mating types. Haploid phase is represented by ascospore along in the life cycle e.g. *Saccharomycodes ludwigii*.

**Sexual Compatibility**

We have already diccussed the various methods by which the two compatible nuclei are brought together in the same cell as a prerequisite stage prior to nuclear fusion (karyogamy). To summrise the process we will consider them groupwise.

In lower fungi, like Mastigomycotima the gametes may be one celled individuals (*Olpidiopsis*) swarm cells of the same (Chytridiales) or different - size (*Allomyces*) or a motile sperm and a non motile egg (*Monoblepharis*) or an oospore and the contents of the anthredium (Oomycetes) or gametangia which are developed at the ends of hyphae as zygophore (Zygomycotina).

Among Ascomycotina, some of the Endomycetaceae have one or multinucleate gametangia and in certain yeasts there is fusion between one celled individuals or in *Saccharomycodes ludwigii* fusion between ascospores. Some members of Pezizales (*Pyronema*) and Laboulbeniales have trichogyne - spermatia conjugation, but Laboulbeniales some times have non motile sperms. In *Phyllachora actinodephne* and *loculoascomycetes* (*Tryblidiella rufula*) results zygote formation due to spontaneous heterokaryosis while highly evolved members of loculoascomycetes (*Bagnisiella indica*) exhibit total absence of sexual fusion since fungi is dikaryotic and zygote like cells originate as heterokaryotic conocytic cells that leads to initiate apothecid pseudothecial compartments. In most ascomycetes the nuclear fusion followed by meiosis take place in the penultimate cell of ascogenous hyphae (ascus mother cell) resulted after crozier formation.

Among Basidimycotina, the dikaryophase uredinales arise by the spermatization of a receptive hyphae (flexuos hyphae), that of hymenomycetes by oidia or majority exhibit simple hyphal fusion between haploid mycelia (primary mycelia) to form dikaryotic vegetative mycelium (secondary mycelium).

**Sex organs in fungi:** Although compatibility is certainly closely related to sex because it governs sexual reproduction, one should not compare it with sex. There are many examples in many fungi that produce clearly distinguishable male and female sex organs (gametangia) on the same thallus. But single individuals (a thallus that has originated from a single spore) of the fungi are in majority sexually self sterile because their male organs are incompatible with their female organs and no plasmogamy can take place. On the basis of the sex, most fungi are classified into three categories:

Monoecious - (Hermaphrodites): In which each thallus bears both male and female sex organs that may or may not be compatible.

*Dioecious:* In which some thalli bear only male and some bear only female sex organs. Very few dioecious fungi have been discovered.

Sexually undifferentiated or indeterminate: In which sexually functional structures are produced that are morphologically undifferentiated (similar) as male or female, perhaps majority of fungi exhibit this kind of sexual process.

• Self fertility in monoecious thallus has the disadvantage of limiting the number of recombinations of alleles during sexual fusion and meiosis, so that there is less genetic variability in the progeny that if nuclei from different thalli has fused i.e. inbreeding is avoided and outcrossing is favoured. Genetic variation is often advantageous to organisms since this is the basis for the evolution by natural selection. And a set of greater likehood of fertilization in monoecious organisms and genetic stability is also advantageous to an organism which occupies a special stable environment. In majority of fungi this disadvantage of self sterility is overcome by their being self sterile. Self sterility is acheived because the thalli are dioecious or opposite mating types occur on the different thalli or in addition self sterility may occur in monoecious or dioecious or sexually undifferentiated thalli in which inbreeding is controlled by one or more genetic factors for compatibility. On the basis of compatibility, fungi are either homothallic or heterothallic or secondary homothallic.

**Homothallism or heterothallism**

Homothallic fungi: In this group every thallus is self fertile and reproduces sexually by itself without aid of another thallus, thus no dioecious fungi are homothallic. In short every individual is self fertile and self compatible, in addition inbreeding leads to genetic homogenity.

Heterothallic (outcrossing) fungi: In this group every thallus is sexually self sterile regardless of whether or not it is monoecious and requires aid of another compatible thallus of a different mating type for sexual reproduction. Since heterothallic fungus requires a compatible partner for reproduction, therefore outbreeding is obligatory. Two fundamentally different types of heterothallism exist: **Morphological heterothallism** - morphologically dissimilar sexual organs are produced by different thalli e.g. Mastigomycotina (*Achlya*); and **Physiological heterothallism** - Which depends on the genetic factors conferring compatibility and may lack differentiated sexual organs (Hymenomycetes - hyphal fusion) or fungus may have morphologically distinct gametangia or fungus may have male and female organs on the same thallus but for diplodization it has basic requirement for genetically different nuclei e.g. *Ascobolus stercorarius, Neurospora* and uredinales (rust fungi). Fertilization in these fungi occur between a male gametangium borne on one thallus and female gametangium borne on another thallus (individual) and further more the thalli must be genetically compatible. Physiological heterothallism is under genetic control and this genetic control is exerted either by two alleles at the single locus (Dimictic) or by multiple allelic series at the one or two or three loci (Diphoromictic) (Table 5.1).

Dimictic - (Two allele heterothallism): In this type, two types of complementory allels control the mating pattern (Dimixis) which is basic heteromictic system in the fungi belonging to Zygomycotina, Oomycetes and Hemiascomycetes. Here single gene with two different compatible mating alleles called as +/-, A/a, $A_1/A_2$, +0/–0 is generally designated.

Diphoromictic multiple allele heterothallism: In this second category there are several alleles involved which are necessary for mating, among which only two compatible alleles must be complementory. These complementory alleles may be at single locus or at two loci or at three loci. Each allele can exist from several to many genes. This category is further divided upto three subtypes depending on whether these complementory alleles are controlled by one make factor, since these exist at single locus or two mating factors existing at two loci or at three loci. Each allele can exist from several to many genes. This category is further divided upto three subtypes depending on whether these complementory alleles are controlled by one make factor, since these exist at single locus or two mating factors existing at two loci or three mating factors existing at three loci, the last one is exhibited only in *Psathyrella coprobia*. These sub types are referred as:

| | | |
|---|---|---|
| *Bipolar multiple allele heterothallism* | = | *Diphoromictic unifactorial heterothallism* |
| *Tetrapolar multiple allele heterothallism* | = | *Diphoromictic bifactorial heterothallism* |
| *Octapolar multiple allele heterothallism* | = | *Diphoromictic trifactorial heterothallism* |

**Bipolar incompatible systems-Bipolar multiple allele heterothallism**

*Diphoromictic unifactorial type heterothallism:* There is a single locus or one mating factor with multiple allelic series occuring at that locus. A cross is compatible if alleles in the mating thalli are different and 50% mating compatibility balances. If we designate the locus as "A", there might be multiple alleles A1, A2, A3 .... An. A thallus carrying the gene A1 can not mate with a thallus also carrying A1, but it will mate with the thallus containing any of the remaining alleles. 37% of the Basidiomycotina exhibit the type of mating system.

**Tetrapolar incompatible system - Tetrapolar Multiple allele heterothallism**

Diphoromictic Bifactorial type heterothallism: There are two loci or two mating factors with multiple allelic series occuring, at each locus. A cross is compatible if both the alleles differ in both the loci and with 25% mating compatibility balance. If we designate two loci, A and B, these loci segregate independently at the meiosis.

Each locus is multiple alleles with at least 100 alleles. The mycellium bearing the alleles $A_1B_1$ will not mate with $A_1B_1$ but will mate with $A_2B_2$ because the alleles at both the loci are different. A cross between $A_1$,$B_1$ is not fully compatible with $A_1B_2$ or $A_2B_1$ because either the alleles at A locus are common or the alleles at the B locus are common.

The research with heterokaryons of *Schizophyllum communae* exhibit common A alleles, prevent the appearance of the clamp connections, disrupt the irregularity of the nuclear distribution and cause morphological and metabolic abnormalities in the mycelium.

Common B alleles prevent nuclear migration, completion of clamp connections, karyogamy and meiosis. Tetrapolarity is exhibited in 63% Basidiomycetes belonging to Hymenomycetes, Gasteromycetes and Smut fungus *Ustilago maydis*.

**Octapolar incompatible systems - Octapolar Multiple allele heterothallism**

Diphoromictic trifactorial type heterothallism: This system is seen in *Psathyrella coprobia* with $A_1B_1C_1$ and $A_2B_2C_2$ with 12.5% mating compatibility balance *i.e.* haploid individual can mate only with 12.5% of the whole population. In this it was found that all three factors regulate nuclear migration and fertility only by two factors.

Secondary homothallic (inbreeding) fungi: In some bipolar heterothallic fungi an interesting mechanism operates, where by two nuclei of opposite mating type are incorporated regularly in each spore. Each spore upon germination gives rise to a thallus containing both $A_1$ and $A_2$ nuclei and consequently behaves as if it were homothallic. This condition is referred as secondary homothallism or homoheteromictic, and fungi with heterokaryotic ascospores or basidiospores. These conditions are either sporadically or accidentally resulted due to sudden environmental changes or due to error in the cellular processes or at the time of ascus formation or may be the results of ascospores developing with nuclei of both mating types. This condition continues and heterokaryons which develop form such ascospores are capable of developing fruiting bodies. Situation exists with two spored basidia in *Coprinus bisporus* or even in 4 spored basidia of *Mycocalia denudata* while in case *Neurospora tetrasperma* and *Podospora anserina*, 4 spored asci with heterokaryotic spores in respect to their mating types.

Investigating the origin and consequences of adopting a new compatibility system is essential, as ecologically related taxa differ in their compatibility systems. A careful examination of the closely allied taxa might indicate correlation between their compatibility system because such correlation has been detected in flowering plants too. Some correlation does exists in a very broad way by comparing the distribution of the compatibility systems with allied group of fungi. Table shows compatibility system in the different groups of fungi and table shows examples of related species of fungi which differ in their compatibility systems. Such systems are important elements of recombination system, but they are only a component of the genetic system upon which selection acts.

**Table 5.1:** Examples of related species which differ in their mating systems

| | |
|---|---|
| *Rhizopus nigiricans* | dimictic |
| *Rhizopus sexualis* | homomictic |
| *Neurospora sitophila* | dimictic |
| *N. crassa* | dimictic |
| *N. tetrasperma* | homoheterodiphoromictic |
| *N. gatapagoscensis* | homomictic |
| *Coprinus ephemerus* | unifactorial diphoromictic |
| *C. congregatous* | unifactorial diphoromictic |
| *C. sassi* | bifactorial homoheterodiphoromictic |
| *C. bisporus* | homoheteromictic |
| *C. stellatus* | homomictic |
| *Sistotrema brinkmannii-I* | homomictic |

| | |
|---|---|
| *Sistotrema brinkmannii-II* | unifactorial diphoromicitic |
| *Sistotrema brinkmannii-III* | bifactorial diphoromictic |
| *Mycocalia duriaeana* | homomictic |
| *Mycocalia denudata* | unifactorial diphoromictic or facultative homoheterodiphoromictic |

**Homogenic and heterogenic**

Incompatiblity: In the past, sexual incompatibility was defined from a genetic point of view, that genetically like gamates or nuclei could not undergo karyogamy. This basic principle was valid for all different incompatibility mechanisms found in higher principle was valid for all different incompatibility mechanisms found in higher plants as well as fungi. During the last decade, however, it has become evident that this is not the only general principle leading to incompatibility. In the study of different geogrpahical races of the ascomycetes *Podospora anserina* (family Sordariaceae), it was found that the pre-requisite for incompatibility can also be a heterogenic structure of the gametic nuclei. Esser and Raper (1965) coined the ferm **Homogenic incompatibility** and **Heterogenic incompatibility** to delimit two restricting systems for sexual propogation on the basis of their genetic determination which serves contrasting but complementory role in the life cycle of fungi. <<Homogenic incompatibility>> promotes zygote formation between individuals with unlike mating type alleles and results in outcrossing between genetically different lines or mating is prevented between strains having some factor. In other words inbreeding is prevented and outbreeding is encouraged <<Heterogenic incompatibility>> might be either sexual inconsequence preventing zygote formation or mating is prevented between strains having different factors so that outcrossing is prevented and inbreeding is encourged i.e. by preventing somatic (vegetative) integration between genetically non-alike individuals, which is also referred as **non-self recognition mechanism**. Both vegetative and sexual incompatibility are different and separate from each other, but both mechanisms are essential for completing the life cycle of the higher fungi.

The efficiency of sexual propagation is governed by recombination of the genetic materials. As recombination of genetically unlike nuclei increases, then fertilization between genetically like nuclei diminishes. This restriction of inbreeding is favoured by dioecy (homogenic incompatibility). The same effect can not be obtained in monoecious forms by homogenics of this system in which genetically like nuclei are incompatible. Homogenic incompatibility is considered as a system diminishes inbreeding and enhances outbreeding.

Heterogenic incompatibility has just a contrary effect in the evolution as this system is based on the heterogenity of the gametic nuclei, it never occurs in a single race. This system enhances inbreeding and diminishes outbreeding and is considered as an isolating mechanism. Here in heterogenics, mutation, which is a result of the recombination, is very slightly transformed from one race to the another race and hence race is the smallest unit of evolution.

**The basic incompatibility reactions:** The various patterns of the incompatibility reactions in the different species of the fungi are based on the type of mating form in a single tetrad. The four products of the meiosis are isolated together as tetrad or its equivalent. Analysis of a large and random sample of the segregants provides information concerning the number, location and linkage relationships of the segregating genes. In all situations, the larger the number of the progeny scored, the more accurate is the information derived.

**Tetrad analysis:** In virtually all members of Ascomycotina and Basidiomycotina, the immediate products of the meiosis are retained as they mature either within or upon the terminal cell (the ascus or the basidium) in which diploid nucleus which is established by nuclear fusion, has undergone meiosis. Meiotic product of tetrad is either discharge forcibly in a group or the product can be removed by mechanical means. Isolation of the tetrad is often easy *e.g.*:

(i) With large basidia and basidiospores, the four spores will adhere electrostatically to a fine needle brought towards them in case of the species of *Coprinus*.

(ii) With smaller basidia and non discharging asci, it is done by simple effective and radially constructed micromanipulator.

(iii) Tetrads of the species belonging to *Ascobolus immersus* or *Neurospora crassa*–which discharge their ascospores violently, can be collected readialy on a slide held at an approximate high above the ascospores.

(iv) Tetrad isolation in the *Saccharomyces cerevisiae* is done either by first separating an ascus cell and then dissolving it's walls with snail gut enzymes or preincubation on the nutrient medium followed by mechanical separation.

In many ascomycetes with linear asci, ascospores can be dissected in the same order in which they are arranged in the ascus so that the order of arrangement of ascospores is related to the orientation of the meiotic and subsequently the mitotic spindle (**ordered tetrads**) e.g. *Neurospora* whose asci contain eight ascospores, each one genetically being duplicated as a result of the further post-meiotic mitosis. The Baker yeast and some four spored asci are linear but a majority of other yeasts and ascomycetes are either oval or clavate, so that certanity of the spore order is not tight. But even those exhibit ordered tetrads. The four spores basidia of the rusts and smuts are linear tetrads.

In other type neither the spore nor their position provides reliable information concerning the sequence of the meiotic events in the tetrad (**unordered tetrads**).

## Other Reproductive Systems

**Heterokaryosis:** Hetrokaryosis is a term to describe precisely the condition of a cell containing two or more genetically different nuclei. (Hansen and Smith 1932). Though dikaryon in the secondary mycelium of basidiomycotina resembles a heterokaryon, it differs because in the dikaryon, nuclei divides at the same rate and are uniformly distributed throughout the mycelium. The degree of coordination does not exist in heterokaryon, instead their nuclei are randomly distributed and vary greatly in the relative number.

Two important features of heterokaryons are:

(i) Nuclei are complementory to each other, express same type of dominant/recessive relationship as found in diploid cells.; (ii) Ratio of nuclear types vary within wide limits and are altered potentially by the environmental conditions.Heterokaryosis enables haploid fungi to shield recessive genes from genetic consitution in response to environmental selection pressure.

This kind of feature is not found in any other kind of living organism. Heterokaryosis is not a substitute for sex, but it is merely a mechanism for juggling whole nuclei so that several nuclei are processed within a common cytoplasm.

Heterokaryosis can arise in following ways:

(i) Mutation of the genes within one of the existing nuclei which automatically establishes a heterokaryon.; (ii) Genetically different nuclei through hyphal fusion, introduced into homokaryon.; (iii) Hyphal fusion between genetically different colonies, which rarely occurs in the lower fungi. In the last two cases, functional heterokaryotic condition is fully established when genetically different nuclei appear in an apical cell and then they can proliferate this condition throughout all subsequently formed hypae.; (iv) Mutation of multinucleate homokaryon coenocytic fungi. The mutant nuclei can subsequently survive, multiply and spread among the wild type nuclei.; (v) Inclusion of the non identical nuclei in a single spore subsequent to mitotic division.

This method allows heterokaryotic association maintained in several sexual generation cycles. (e.g. *Podospora anserina, Neurospora tetrasperma*), while in asexual reporduction, this is accomplished through the formation of the multinucleate asexual spores.

Heterokaryosis plays a key role in the variation among fungi imperfecti and phenotypically the heterokaryons may be quite different from their potential types, but exert certain amount of control over the common cytoplasm and also carries genetic determinants.

Once a heterokaryotic branch arises near the colony margin, it can further branch and give rise to homokaryotic sector. This section from the rest of the colony in its growth, colour, density of the sporulation or alternatively this section may go unnoticed on agar plate, because they differ either in their pathogenicity or in their biochemical activities.

Another mode of heterokaryon breakdown is automatically during asexual sporulation, provided spores are uninucleate as in *Aspergillus nidulans*. The situation is more complex with multinucleate spore, if all their nuclei arise from a single nucleus in the spore mother cell as in *Fusarium*, then spores developed are homokaryotic, but in the *Neurospora* several genetically different nuclei enter the developing conidia, so that conidia may be homokaryotic or heterokaryotic. Heterokaryosis is a way in which virulent strains aquire virulence. e.g. *Thanetophorus cucumeris*.

Thus heterokaryosis is an alternative mode of reproductive system in which fungi have potential for continuous variation in the response to the environment. It enables recessive genes to be shielded from selection pressure. Some fungi have ability to revert to homokaryotic condition throughout most of their life cycle (at least in their asexual reproductive stage). All these features make heterokaryosis a striking biological phenomenon. The occurrence of heterokaryosis in the nature is still unclear, most of the work is based on the artificial heterokaryons obtained from parental homokaryons with auxotrophic (nutritional) pressure for heterokaryosis.

Heterokaryons from virulent strains are easily obtained by spontaneous and induced way of crossing between a virulent and a non-virulent strain *e.g.* hetorokaryosis gives rise to races of strains of the rust fungi. In well documented examples in case of *Puccinia striformis*. Thus heterokaryosis is considered as the first significant step towards the parasexual process.

**Parasexual Cycle:** Parasexuality was discovered by Pontecorvo (1956) during his early studies on the heterokaryosis in *Aspergillus nidulans*. Pontecorvo and his colleagues showed that three are four septate and largely independent stages involved:

(i) Fusion of adjacent somatic hyphae and exchange of nuclei and establishing a heterokaryon; (ii) Fusion of different nuclei in the vegetative hyphae to form a somatic diploid. Once diploid nucleus is formed, it is relatively stable.; (iii) Somatic recombination in which mitotic crossing over occurs within heterozygous diploid nucleus, it is of rare occurrence and at the most once on each chromosome. Several crossing over points occur on each chromosome during meiosis: (iv) Non meiotic reduction of the altered diploid nuclei to the haploid ones. e.g. initially diploid nucleus gives rise to nuclei with 2n + 1 or 2n – 1 chromosomes (*Aneuploid* - nucleus with incomplete multiples of the haploid) that are unstrand and tend to revert to euploid by repeated loss or gain of chromosomes in successive divisions *i.e.* 2n + 1, nucleus to diploid ones and 2n - 1, nucleus to haploid ones. *e.g.* In *Aspergillus-nidulans* (n = 8) nuclei detected are with n = 17 (2n + 1), 15 (2n – 1), 16, 12, 11, 10, 9 and 8 so this leads to a strong support for haploidization.

Parasexuality is a practical means for producing genetic variation, which is wide spread in the conidial fungi like species of *Aspergillus, Acremonium, Fusarium* and *Verticillium*.

Mitotic recombination and developling heterokaryon frequency is increased by **X rays, UV**, mitomycin and nitrous acid. Even with the help of the low concentration of p - flurophenylamine benlate stimulate haplodization, so that fungus undergoes parasexual cycle. Parasexuality is being described in sexual fungi, it is not found in *Neurospora crassa* or *Podospora-anserina* due to their genetic basis to undergo normal sexual reproduction. Several machanisms (including parasexuality) are proposed to explain the origin of new races in rust like *Puccinia graminis* and *Melampsora lini*. In such cases somatic hybridization term has been used to described phenomena for parasexual recombination. In Deuteromycotina which like a heterokaryon during their normal vegetative growth, parasexuality enables them to undergo genetic recombination. Due to this ability, members of fungi imperfecti are widely employed in stain improvement programme. Parasexuality is used to determine linkage groups, linkage analysis, order of genes and position of the centromeres. Genetic recombination is achieved in parasexual cycle is on much smaller scale than in sexual cycle during meiosis. It is possible to breed asexual organism used in fermentation so as to combine in one strain, the desirable properties obtained from different strains.

The genetic information from two genotypes is brought together into new genotype through genetic recombination. It is effective means of increasing the genetic variability of a population. In mutant screening (which is done either by random or empherical section) each high yielding mutant is derived from wild strain after a series of mutant screening steps. In short cell lines with 10 mutants contains 10 new genotypes. By corssing the last high yielding mutant with that of the original wild strain, different genotypes are elicited. This is why crossing is must in high yielding mutant strains from two different lines. The advantages of the genetic recombination include:

(i) Different alleles of the present strain with increased secondary metabolites production are brought together in one strain so that the culminative effect of such mutations is greater than the effect of the single is mutation.; (ii) In course of strain development programme, there increase in the yield after each stage of mutation. Mutation which are selectively enriched due to their increased productivity, there is development of inapparent mutation which prevent further increase in the metabolites production through pleiotrophic influence. With favourable crosses these unfavourable alleles are replaced by alleles of one of the parents in the cross. (iii) High yielding strains actually increase the cost of the termination, because of the changes in the physiological properties, like greater foaming and requirement of the culture medium. By crossing back with wild type strain, high yield strains plus fermentation properties are regained.

In case of the fungi belonging to Deuteromycotina, used in the fermentation industry, genetic recombination processes are employed to produe strains with increased penicillin titers either by heterozygotic diploid obtained by mutation screening from parent strain or by haploid recombinants to produce stable strains with improved antibiotic production.

Extensive efforts have been made to establish protoplast fusion in the species of *Aspergillus, Penicillium* and *Cephalosporium*. In case of *Cephalosporium-acremonium*, strains obtained after protoplast fusion. There is increase in the Cephalosporin production, culture sportulates well and increase in the growth rate because there is increase in the formation of the heterokaryons. While Chang *et al.*, 1982 used fusion of two protoplasts to combine strains of *Penicillium chrysogenum*, on synthesizing **Penicillin V** and adverse morphology, other desirable morphology with significant quantities of **Penicillin OH, V** and desired morphology.

Interspecific hybridization allows genetic information from desired species to be combined **in vivo** in order to create new modified secondary metabolite product. This technique is utilised in examining different antibiotic manufactured.

Heterokaryosis and parasexuality are of probably universal occurrence among filamentous Ascomycotina and Basidiomycotina, but not known to occur in the Mastigomycotina. These cycles provide for direct somatic variation in the vegetative phase of the life cycle but additional and significant variation is incorporated through the sexual reproduction. Heterokaryosis has been considered as a probable mechanism leading to establishment of sex in fungi.

# 6. MYXOMYCETES

De Bary (1887) considered myxomycetes to be nearly related to members of Animal Kingdom. Botanists looked upon their mature fruiting bodies as fungi and gave them names, many of which are still maintained. These organisms are naked during all stages of development except the culminating spore stage. This naked stage is more or less amoeboid and ingest particles of food. At germination of the spore the escaping naked protoplast may be simple uninucleate amoeboid cell or myxamoeba, or it may have one or two anterior flagella which eventually are retraced, leaving the cell a myxamoeba. These are capable of multiplication by fission. Later by the division it becomes multinuclear plasmodia, or these may arise by the fusion of separate myxamoebae as well as by growth and nuclear division.

Organisms are usually classified but the zoologists as belonging to Phylum Protozoa, Class Sarcodina. This classification is quite different from that mostly used by the botanist.

Myxomycetes or myxogasters or slime molds compose the most numerous group of this subclass. They are terrestrial organisms or inhabitants of manure, decaying wood, bark, fungi etc. Their spores are produced upon or with in aerial sporangia and more or less dependant upon wind for their dissemination.

They are world wide in distribution occurring in tropical, subtropical and the temperate zones. While some species occur all over the world. There are others that have a restricted distribution. Nearly 450 species of Myxomycetes are known up to date. Moisture and temperature play a significant role in their development, occurrence and distribution. They are most abundant in rainy season although months and there are still others that can be encountered during winter also.

They are commonly known as true slime molds or myxomycetes. These organisms have following life cycle stages.

(1) Three types of uninucleate cells one of which is flagellate.

(2) A multinucleate somatic phase known as plasmodium.

(3) Resistant stage consisting of sclerotium.

(4) Reproductive stage that culminates in the production of stationary sporophores containing walled spores that may be dispersed by wind and water or by arthropods that live and feed on the sporophores. Myxomycetes have two types of mitotic divisions. (a) centric (b) acentric.

They are a cosmopolitan group and found on different habitats including even well manicured lawns and flower beds, and on wood bark. They occur on decaying logs, dead leaves and other organic water. They are widely distributed. About 500 species are known, and of these about 200 are in the order Physarales which have large plasmodia. The amoebae of *Physarum polycephalum* feed on bacteria, and *Dictyostelium* are readily grown in two membered culture with *E. coli.* They have been grown in pure culture on a complex medium, but multiply more slowly. The nucleus has a prominent nucleolus, a food vacoule and a contractile vacoule are present. If a culture is flooded with water, the amoebae elongate and turn into flagellates. The flagella are smooth and emerge from the anterior end of the cell. There are usually two flagella, one pointing forward is active the other points backward, is held close to the surface, and is inactive. The flagellates neither feed nor undergo cell division and when free water disappears revert to the amoeboid state.

Slime molds are also common on dung, only a few of these may be restricted to dung substrates. Another specialized niche appear to be dead branches attached to living trees. Plasmodia and sporophores of certain species of myxomycetes often appear in lawns and on some ornamental plants. *Physarum* is the most abundantly present myxomycetes may form colonies several feet in diameter on city lawns. Moisture and temperature also are important abiotic factors and contribute to the periodic appearance and seasonal nature of some species. During a rainy season myxomycetes begin to appear as a group in May in Northern area in Inida through October. Some are more abundant in spring, some in the middle of summer and others in early fall.

Spores of myxomycetes are liberated from their sporophores by a variety of factors including wind, water and the activities of animals. Spores appears to be extremely resistant to unfavorable conditions.

Most common genera are– *Arcyria, Trichia, Metatrichia, Ceratiomyxia* and *Fuligo*.

When all the bacteria present are consumed, the amoebae turn into thick walled cysts. Which can survive for a long period in the absence of nutrients. In the presnece of bacteria a cyst will germinate and an amoebae emerge. The amoeboflagellate phase is able to multiply in the presence of bacteria swim when flooding occurs, and survive periods without nutrients.

**Plasmodial phase:** It usually requires mating between two genetically different strains. Plasmodium initiation occurs with the highest frequency if the two strains differ at two loci, designated mat A and mat B. At each locus many alleles have been found so the number of possible mating types is large. If two amoebae differ at the mat B locus the probability that they will fuse to give a diploid cell is greatly increased. Diploid cell or zygote develop into a plasmodium is greatly increased if it contains two different mat A alleles. *Physarum polycephalum* is self sterile (heterothallic) and has diploid plasmodia since zygote production is involved in their origin. In these a mutation at the mat A locus enables amoebae to give rise, without mating to plasmodia which are haploid. The amoebae of these strains can also mate with other strains to produce diploid plasmodia. Studies on other myxomycetes have demonstrated self fertile (homothallic) species, in which mating occurs within a clone. The sexual behaviour of myxomycetes is thus very flexible. Not only they are self sterile, and apomictic species, but all three forms of behaviour have been found among strains of a single species, *Didymium iridis*.

Zygote grows and under goes mitosis without cell division occurring. Nuclear division is synchronous young plasmoduim has in turn 2, 4, 8, 16 nuclei. Uncountably large numbers are soon reached and plasmodia may cover many square centimeters and contain millions of nuclei. Nuclear membrance remains intact. Plasmodia increase in size by fusion with each other as well as by growth. As the size of plasmodia increase protoplasmic streaming becomes more pronounced and ultimately develops into the shuttle remaining in well defined channels characteristic of the mature plasmodium. Plasmodia can be grown in pure culture on a soluble medium containing a suitable carbohydrate, nitrogen source, mineral salts and vitamins either on an agar medium or in shaken liquid culture. In nature, plasmodia can surround and digest liquid culture, however shows that plasmodia are capable of absorbing nutrients and in nature plasmodial nutrition is probably partly ingestive and partly absorptive.

The plasmodium as a result of its size, is able to move further and attack for larger organisms than can amoebae. Plasmodia are surrounded by a mucopolysaccharide slime sheath gives some protection from desiccation and aids locomotion.

When genetically identical plasmodia meet they fuse to form a single plasmodium. If however the plasmodia belong to different strains, then fusion may not occur, or if it does, the nuclei of one strain may be eliminated sometimes with considerable destruction of protoplasm.

If plasmodia is starved and light is absent than plasmodium becomes a Sclerotium, which consists of numerous spherules each thick walled and containing several nuclei and protoplasm, it can survive for many years. When nutrition is present the spherules germinate and the protoplasts merge to form a plasmodium. A starved plasmodium merge to form a plasmodium. A starved plasmodium in light gives rise to a fruit body containing spores. Meiosis occurs during sporulation to return the organism to the haploid state. Spores are capable of prolonged survival and are readily dispersed if they arrive at a suitable site germinate to release amoebae.

**Classification**

Martin (1960) has given the following scheme of the classification for the class myxomycetes.

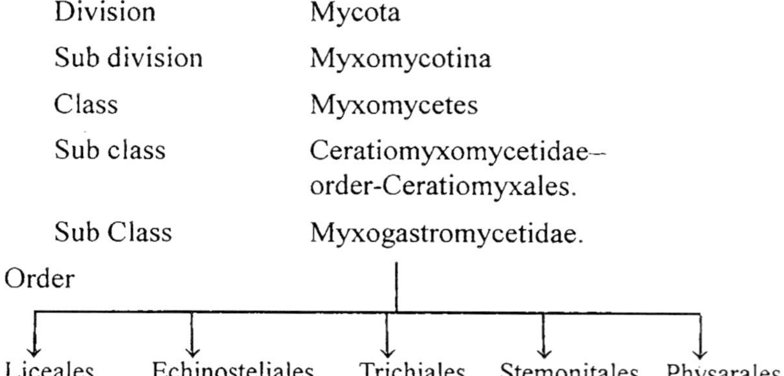


**Economic importance:** Myxomycetes are holozoic-ingesting bacteria, protozoa and food particles and are secondarily saprobic, absorbing nutrient in solution. They thrive on dead and decaying organic matter such as bark, wood, things, leaves etc. Such decaying organic matter not only rich in other microflora, some take active part in decay, but also contains inert food in both soluble and insoluble form as a result of such decay.

**Plasmodium:** It represents the vegetative phase of the Myxomycetes. It resembles a protozoan and may be regarded as a gigantic amoeba. It is a naked mass of protoplasm surrounded by a very thin plasma membrane.

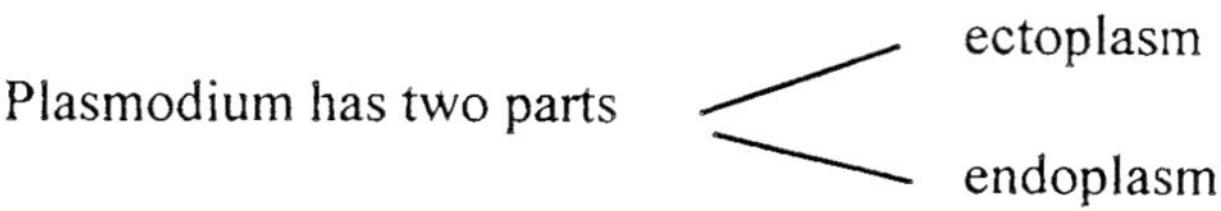


Ectoplasm is the outer hyaline thin plasma membrane and endoplasm is the bulk of internal protoplasm. It moves on the substratum and feeds like an amoeba. It spreads itself in the form of various veins which cover a large area or the whole of the substratum. Plasmodia are holozoic and saprobic in nature.

Plasmodium is the active streaming of its protoplasm. The protoplasm first moves in one direction at an accelerated rate till it reaches a maximum value, after which the streaming slows down and ultimately comes to a standstill for a movement. Then the protoplasm again starts to flow, but this time in opposite direction. Again the rate increases to a maximum and the whole cycle is repeated.

Plasmodia are brightly coloured, but the white, yellow and brown colours predominate. Alexopoulos (1960) described the various types of plasmodia.

**Phaneroplasmodium:** It is the most massive type of plasmodium known among the myxomycetes. It is marked by (1) granular nature of its protoplasm, which makes it easily observable at an early stage. (2) The conspicuous differentiation of its thick veins into streaming endoplasm and the stationary, jellified ectoplasm. (3) The thick fleshy fan which when fully formed, spreads as a perforated or solid sheet of protoplasm with many channels in which the endoplasm streams.

**Aphanoplasmodium:** It is found in *Stemonitis* and *Comatrichia*. It is marked by (1) the non-granular nature of its protoplasm, which render it difficult to observe until it is fully formed. (2) absence of jellified ectoplasm, except in the largest veins. (3) nature of its advancing fan which often consists of an open network with very delicate strands.

**Protoplasmodium:** It remains microscopic through out until it fruits and give rise to single sporangium. Its protoplasm is granular, but it exhibits no advancing fan, no channels, no veins, no reticulation, no rapid, reversible streaming. Its protoplasmic streaming is sluggish and irregular can be detected under high magnification. It is characteristic of the order Echinosteliales.

**Intermediate type:** It occurs in Trichiales. This has something in common with all the other three types. It is intermediate between phaneroplasmodium and aphanoplasmodium. It grows as fans and has abundant dense granules, thereby being readily visible. It has no stationary outer gelled layer in the veins except the largest ones and all the protoplasm within the channels moves in the reversible flow. It feature with protoplsmodium is its random movement of its granules in the protoplasmic sheet.

**Fructifications:** Vegetative phase of Myxomycetes is fruit body or sporophore. Spores are produced exogenously on the surface of a sporophore (*Ceratiomyxa*). Plasmodium gives rise to columnar pillars which bear spores on their external surface. These pillars may be simple or branched and usually gregarious to densely crowded into small or large patches. Cottony growth is observed on decaying logs often consists of the fructification of *Ceratiomyxa*. In rest of the myxomycetes fructifications are always endosporous hence they comprise the sub class Endosporaceae. Spores are produced inside the fructifications or in other words. Spore mass becomes covered by a periodical membrane. Three type of fructification has been reported in endosporous fructifications:-

**Plasmodiocarp:** It stimulates a plasmodium and hence the name plasmodiocarp. The protoplasm of a plasmodium accumulates in a few large veins where the spores are delimited and surrounded by a peridium. Plasmodiocarp is continuous branched, vein like and netted as in *Hemitrichia serpula*.

Plasmodiocarps are shorter, discontinuous and discrete but still elongated, straight to curved, crescent horse shoe shaped, simple or branched. In some case plasmodiocarps are reduced to small, spherical, cushion shaped or pulvinate sporangial types.

**Sporangium:** This is the most frequent type of fructification formed in the Myxomycetes. In this type the plasmodium segregates itself into numerous small discrete units, each of which develops into a sessile or stalked sporangium. Sporangia are usually crowded and packed into small to large patches or clusters. Sporangia may be spherical to ovoid to fusiform to cylindrical, small to large and differently coloured. Spherical sporangia may be depressed globose to discoid to lenticular– *Physarum compresium*.

**Aethalium:** Whole of the protoplasm of a plasmodium, is incorporated into one usually, large, cushion shaped, fructification, surrounded by a thick periodical membrane. They are less commonly observed, but in *Fuligo, Lycogala* and *Reticularia* it commonly occurs.

**Spores:** The spores of myxomycetes are always one celled and monotonously globose to sub globose throughout. Thus the shapes of the spores seldom contributes to the taxonomy of these fungi. Spores are ovoid or some what ellipsoid in *Ceratiomyxa fruticulosa, Fuligo cinerea.* In *Diderma,* globose, sub globose, ovoid, and ellipsoid spores all appear to be equally represented in the some sporangium. Spores are less than 5μ or more than 15μ in diameter. Spore marking may be verrucose or reticulate. Spores are known to be perfectly smooth walled–*Ceratiomyxa*, nearly smooth in some species such as *Cribraria, Arcyria* and *Stemonitis* (Fig. 6.1 A, B, C, M, N). Spore wall may be very minute or very prominent, spine like or usually blunt. Reticulate type of spore marking may be composed of irregular and incomplete or regular and complete reticulation. Reticulate band may be broad and sometimes pitted or narrow and without pits. Reticulations may be large and small.

Colour of the spores is highly characteristic and is often employed in the segregation of the orders of Endosporae. Spores are black or dark brown in colour in Physarales and Stemonitales. Where as they are bright yellow or brown or reddish in Trichiales and Liceales. Spores are colourless or hyaline in Exosporeae. Spores are produced singly through out the myxomycetes. They may be loosely present or arranged in clusters.

**Dissemination of spores:** The spores of the myxomycetes are undoubtedly disseminated by wind and the capillitium plays an important role in their discharge from the fructifications. After the peridium is split open, the packed mass of spores and capillitium becomes exposed and loosened considerably owing to the expansion of the capillitial threads. In this way the loosened and raised spore mass is fully exposed to the action of the wind. Closely packed elaters are relieved when their ends spring free from the mass, thus throwing the adhering spores in the air.

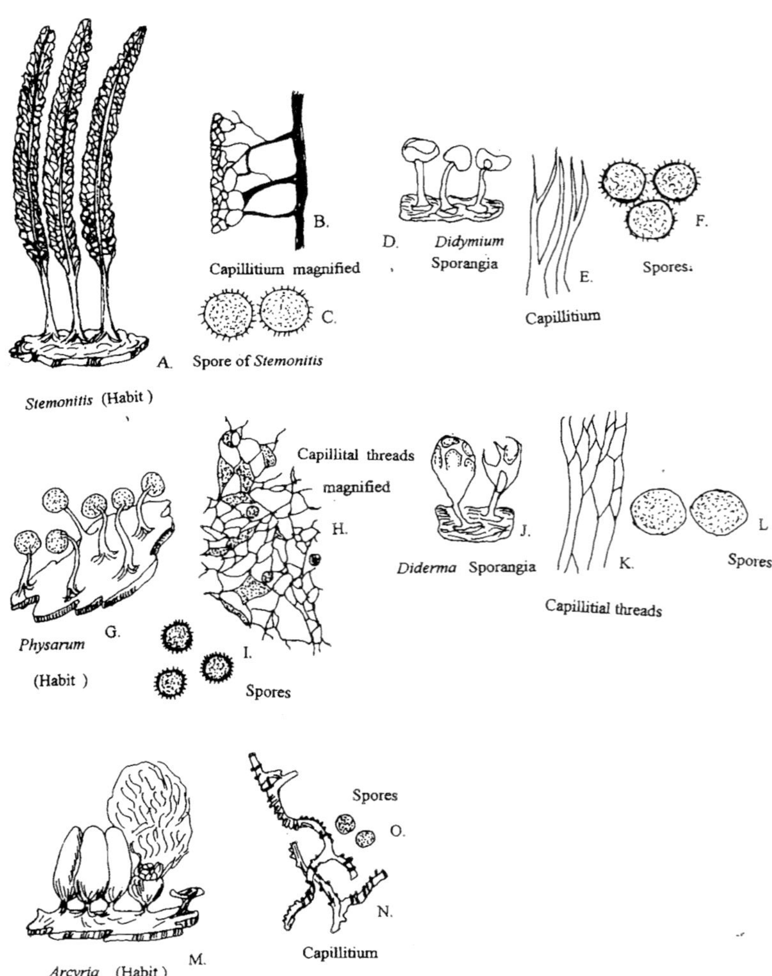


**Fig. 6.1:** Different types of Myxomycetes

**Longevity of spores:** Smith (1929) and Elliot (1949) reported that spores of 21 species were able to germinate even after 5 to 32 years. The nature of resistance may be mechanical because of the thickness of the spore wall, or may be protoplasmic. Percentage of spore germination will go on decreasing with age.

The presence or absence of slime in one or more of the spore wall, columella, capillitum, peridium, stalk and hypothallus is well known among different myxomycetes, forms one of the important bases for the segregation of orders and families. Boic (1925) made micro chemical tests on the fructifications of a number of species and concluded that frond substance of the spore membranes is pure cellulose and the wall of the peridium is mainly of cellulose, but the capillitium is composed of protein. Ulrich (1943) reported that chemical composition of the membranes of fungi, it appears that chitin, which is present in the plasmodiophorales.

**Sclerotia:** Thick walled spores form the chief resting stage of the Myxomycetes, sclerotia may also sometimes be formed by the plasmodium to tide over adverse environmental conditions. Sclerotia loose their viability with age and this is correlated with water content. The water content of Sclerotia decreases with time and so is their viability. Dormancy of sclerotia can be broken easily. When placed under moist conditions, it at once grows out into a typical plasmodium.

**Reproduction:** Single swarm cells (myxamoebae) giving rise to plasmodia, which later on lead to the formation of fructification. A single swarm cell (myxoamoebae) can multiply into a population of swarm cells which may have fused in pairs before giving rise to a plasmodium. **Sexual reproduction** is the main and normal means of reproduction and multiplication in the Myxomycetes.

**Factors affecting spore germination:** The various factors like age, moisture, temperature, hydrogen ion concentration and nutrients had a marked affect on the germination of spores of myxomycetes. Alternate wetting, drying and fluctuating temperature promote germination. Optimum pH for spore germination ranged from neutral to strongly acidic. Optimum temperature ranged from 22°C to 30°C. Mass sowing of spores gave much better germination than sowing the single spores. Spores of myxomycetes are thick walled and need a thorough wetting and softening of the wall. Time required for spore germination varies with environmental factors, the age and the species.

Spore germination is of two types:

Exosporeae

Endosporeae

In exosporeae germination, there is a general softening of the spore wall and nucleate protoplast emerges from a pore as a globose or slightly amoeboid naked body. After a short while, protoplast enters in the thread stage which is usually maintained for several hours. The protoplast elongates, bends, contracts or expands but shows no amoeboid movement. Thread stage then passes into the globose condition, which under goes cleavage into four uninucleate globose protoplast. Tetrad stage passes into the octant stage, then each portion into two uninucleate globose protoplasts. Protoplast separate, develop flagella and thus there are formed 8 pyriform swarm cells which swim away.

**Endosporeae:** Each spore during germination dehisces by a pore or by a split in the spore wall and the method of dehiscence may be characteristic of a species. A spore on germination, liberates usually one, sometimes 2, 3 or 4 swarm cells of myxamoebae. Emerging protoplast may be a swarm cell from the very beginning or it may be amoeboid to begin with remain quiescent in that condition for a short while and develop flagella and swim away as a swarm cell. On germination, it gives rise to a swarm cell (myxamoebae).

**Multiplication of swarm cells:** Swarm cells multiply by mitotic division and it never divide in the flagellated condition. They loose their flagella, come to rest, become spherical and then divide by mitosis.

**Gametic fusion and formation of plasmodium:** Spores of myxomycetes are haploid. After a period of rest, they germinate and give rise to one or more swarm cells. They multiply by mitotic division. The swarm cells or myxamoebae behave as sexual gametes and fuse in pairs. Plasmogamy is immediately followed by karyogamy. Flagella, if present are dissolved and the zygote soon prepares for the most active stage of the myxomycetes. Immediately it puts forth the pseudopodia and begins to grow by engulfing microbes. It may also engulf swarm cells and other young zygotes. When several zygotes are growing together into young plasmodia, they under go coalescence. Zygote nucleus undergoes repeated mitotic divisions to give rise to a multinucleate plasmodium. The plasmodium may branch forming bigger and smaller veins which often anastomoses, thus forming a netted structure. It crawls over the substratum and represents the most active assimilating stage of the myxomycetes.

Various factors such as nutrition, temperature, light, pH, moisture, aeration, age injury, etc. have been suggested to have a bearing on plasmodia to fructify. Higher the temperature greater the acidity required for the fruiting of this species. Nonpigmented plasmodia fruited equally well in light and darkness. Yellow plasmodium require light for fruiting.

**Meiosis:** Meiosis takes place in the fructification just before spore formation. Each uninucleate sporogenous cell gave rise to a single spore. After the completion of the meiotic division of nuclei the protoplasm of the fructification undergoes cleavage into uninucleate protoplasts. Each uninucleate protoplast becomes spherical and develops a thick wall around it and becomes a spore. It develops colour and ornamentation on its wall, characteristic of its species. Spores are packed closely in between the capillitial threads. Spore mass is dry powder and spores are easily disseminated and liberated on the dehiscence of the peridium.

In *Trichia* capillitial threads are formed as a result of intraprotoplasmic secretions and they arose from the vacoules. In *Didymium* capillitium develops as a result of anastomoses between tubular invaginations from inner periodical wall and the central vacuoles resulting from protoplasmic condensation of maturing sporangia.

***Fuligo:*** It is one of the largest of slime molds. It has a similar capillitium but the spore fruits are united into a single large convolute aethalium.

***Dictydium:*** Sporangia is stalked, has no internal capillitium, but when the peridium disappears it leaves numerous longitudinal ribs that run from base to apex like the lines of the globe.

***Comatrichia:*** It occurs on dead wood, twigs, bark and leaves, sometimes on mosses, lichens and living leaves. Fructification sporangiate stipulate, noncalcareous, sporangia scattered, gregarious, densely crowded, cylindrical, globose or ovoid, stipe jet black, solid, rigid extending into the sporangium as a columella, peridium usually evanescent, some times persistent, capillitium originating from columella, well developed, primary branches, from columella branching and anastomosing to form a loose network, ultimate branch lets free, surface net absent. Spores black, dark, purpule, brown in mass. Six species of *Comatrichia* are reported from India.

***Arcyria:*** It occurs on dead woods, dead leaves, or other plant parts. Fructifications sporangiate, sporangia mostly stipulate, sometimes sessile, stipe often filled with spore like but larger vesicles, peridium fugacious above and persistent below as a cup or calculus, capillitium composed of network of branching and anastomosing tubular threads which are variously marked with half rings, with inconspicuous spiral bands. Capillitium usually strongly elastic, often expanding to more than twice the weight of the sporangium after the disappearance of the upper fagacious periodical portion, the remaining attached to the sides or merely to the base of the calyculus, dehiscence by the disappearance or rupture of the upper fagacious part of the peridium, leaving its lower part as persistent deep up, along a definite line of dehiscence. Spores variously coloured yellow, brown, reddish or greenish in mass, pallid or pale in transmitted light. (Fig. 6.1 M, N).

***Hemitrichia:*** Fructifications mostly sporangiate, sometimes plasmodiocarpous, sporangia sessile or stipulate, peridium tough to cartilaginous, persistent stipe when present, solid or filled with spore like vesicles or amorphous material, capillitium composed of tubular threads which branch and anastomoses into a more or less elastic net, with or without free ends, threads characteristically marked by 2 or more spiral bands, dehiscence irregular, peridium fugaceous in the upper thinner portion, where as its lower portion remaining persistent as an irregular cap.

Spores yellow, yellowish brown, orange or red in mass, pallid by transmitted light, verrucose or reticulate, reticulation more or less raised and making the spores appear double walled. It occurs on dead wood, dead leaves or other plant remains. Four species of *Hemitrichia* has been reported from India.

***H. serpula:*** Spores coarsely and prominently reticulate, fructifications plasmodiocarpous, plasmodiocarp large and netted, capillitium spinulose.

***H. imperialis:*** Spores nearly smooth to verrucose, or reticulate when reticulations are faint. Sporangia cylindrical light copper coloured densely clustered and heaped.

***H. clavata:*** Sporangia pyriform to clavate, not copper coloured, never heaped, stalk short, expanding and merging into the base of sporangium above. Sporangia typically clavate, capillitum minutely roughened.

***H. stipitata:*** Stalk long, cylindrical not expanding and not merging into the base of the sporangium above, sporangia typically pyriform, capillitium smooth.

***Trichia:*** Fructifications sporangiate sometime subplasmodiocarpous, sporangia sessile or stipulate, peridium membranous but firm, persistent, capillitium elastic, composed of short, free, simple or sparsely branched elaters which are characteristically marked by 2–5 or rarely more, spiral bands and are pointed or occuminate at their tips, dehiscence irregular.

Spores yellow, yellowish brown or reddish in mass pallid by transmitted light, versucose or reticulate not smooth. The reticulation is more or less raised and forms a border around the spore. Thus reticulately marked spores mostly appear double walled in a sectional view. Occur on dead wood, dead leaves or other plant material.

From India ten species have been reported *i.e. T. varia, T. lutescens, T. favoginea, T. affinis, T. persimilis, T. decipiens, T. subfusca, T. botrytis, T. floriformis.*

***Physarum:*** Fructifications sporangiate to plasmo-diocarpous, often both inter mixed, rarely pseudoaethaloid, sporangia stipulate or sessile, stipe when present, usually tubular, translucent or opaque and stuffed with lime or dark amorphous material, Peridium mostly single, sometimes double, more or less calcareous, columella present or absent, capillitium typically physaroid, composed of network of calcareous nodes and limbless hyaline, connecting slender internodal threads or tubules, line in the stripe, peridium columella and capillitium is always in the form of amorphous granules and is never crystalline dehiscence usually irregular, rarely specialized *i.e.* locate or stellate or by longitudinal fissure (Fig. 6.1 G, H, I).

Spores black or dark brown in mass violaceous brown by transmitted light.

On dead leaves, twigs, plant debris, rarely dead wood, often on living herbaceous plants.

*Physarum* is the largest genus among all the myxomycetes and there is no end. About 41 species of *Physarum* are recorded from India.

***Diderma:*** It commonly occurs on dead leaves, twigs, wood, mosses and plant debris, sometimes on living plants, Fructifications sporangiate to plasmodiocarpous, sporangia sessile or stipulate. Peridium usually double layered. It is the diagnostic feature of this genus. In most species the peridium is plainly double, two walls being often remote and distinct, they are sometimes closely applied. In few species inner periderm is absent and outer single peridium maintains its calcareous nature. The periodical hence is consistently amorphous and granular in all species of the genus. The peridium have crystalline lime but it is never in the form of stellate crystals. Columella usually conspicuous, sometimes merely represented by the thickened or dome shaped base of the peridium, capillitum usually abundant, delicate or rigia, composed of usually purplish brown, sparsely branched thread, paler at the extremities, capillital thread sometimes marked by nodular thickening, which are devoid of line dehiscence usually irregular, sometimes stellate or circumseissile (Fig. 6.1 J-L).

From India 17 species of this genus has been reported.

***Didymium:*** It commonly occurs on dead leaves twigs, wood, bark, mosses, sometimes on herbaceous living plants, fructification sporangiate to plasmodiocarpous, never aethaloid, sporangia sessile or stipulate, peridium single or double, usually thin and membranous, rarely thick, tough or cartilaginous, periodical lime crystalline, crystals stellate, powdering the surface or united into a crust. Columella usually well developed sometimes reduced to a merely thickened base, capillitium extending from the columella to the peridium, usually abundant, delicate or rigid, composed of usually violaceous brown, sparsely branching and anastomosing, limbless threads, sometimes bearing dark nodular thickening, dehiscence irregular. From India 17 species have been reported (Fig. 6.1. D-F).

**Stemonitales**

***Stemonitis:*** Fructifications sporangiate, stipulate, non calcareous, sporangia gregarious to densely clustered, cylindrical, black or dark coloured, stipe jet black, solid, rigid, extending into the sporangium as a prominent columella, peridium evanescent hypothallus conspicuous, dark coloured or silvery, membranous, often confluent. Capillitium abundant, dark purpule to brown arising from the whole of the columella, primary branches from the later branching and forms a loose network, dehiscence irregular, peridium evanescent, leaving the spore mass enclosed by the surface net only. Spores black or dark brown in mass, purpule or brown by transmitted light. From India 13 species have been reported (Fig. 6.1 A-C).

**Plasmodiophoromycetes**

The plasmodiophoromycetes are obligate endoparasites of vascular plants, algae and fungi, which usually cause an abnormal enlargement of the host cells called hypertrophy, and an abnomal multiplication of host cells called hyperplasia. Due to this enlargement of infected portion of the host and in flowering plants takes place.

Some species parasitize fresh water algae such as *Vaucheria*, or acquatic fungi such as *Saprolegnia, Achlya* and *Pythium*. Some species parasities acquatic and marsh vascular plants in the genera *Isoetes, Halophila* and *Juncus*.

Only two species of Plasmodiophoromycetes are of economic importance: *Plasmodiophora brassicae* and *Spongospora subterranea*.

**General character:** Somatic phase of this class is plasmodium that develops with in the host cells. Two types of plasmodia is there *i.e.* sporangiogenous plasmodia and cystogenous plasmodia. Sporangiogenous plasmodia give rise to thin walled zoosporangia containing zoospores, where as cystogenous plasmodia cleave into thicker walled cysts each of which eventually gives rise typically to one zoospore. The zoospores from both zoosporangia and cysts are similar, each bearing two unequal anterior whiplash flagella. Zoospores arising from cysts were termed primary zoospores, while those arising from zoosporangia were termed secondary zoospores.

Nuclear division in plasmodiophoromycetes is of peculiar type. During division, an intranuclear spindle is formed on which the chromosomes are arranged in a non-continuous ring around the nucleolus. As the chrmosome split, a ring of chromosomes passes to each pole. The nucleolus persists for some time and elongates. The elongated nucleolus surrounded by a chromatin ring appears like a cross when viewed from the side. Akaryotic phase (a = not + *karyon* = nucleus) has been reported in both sporangiogenous and cystogenous plasmodia of some plasmodiophoromycetes shortly after the cessation of the cruciform division.

***Plasmodiophora brassicae:*** Finger and toe, club root disease.

This disease attacks both cultivated and weed plants, belonging to the order cruciferae *i.e.* turnip, cabbage, mustard, rape, wall flower, charlock, shepherd's purse, hedge mustard and others. The disease is wide spread in this country, and has done great damage to crops, especially since the use of acid artificial fertilisers to became common and the use of chalk and lime declined.

**Symptoms:** On sowing, young cabbage plants from seed beds for transplanting the roots sometimes show irregular thickening or knob like swellings. At first the swellings are small but as the plants get older they increase in size. A seedling so affected does not grow normally the upper part develops very slowly. The leaves being small and soon turning yellow, and in the case of cabbage, cauliflower, and broccoli no head is formed. The root is more or less deformed the fancied resemblance to fingers and toes giving to the disease one of its popular names. When a diseased part in its early stages is cut across we find the swelling solid and of greenish colour, with small, while opaque patches visible on the cut surface; at a later stage the swollen parts turn brown and decays. In dry soils the rotten portions become brittle, and fall into powder or small fragments; in damp stiff soils the decayed mass in semi liquid, and of offensive odour. On pulling up a plant at a late stage of the disease a blunt, woody stump may be all the remains of the root. Swedes and turnips generally show nodular growths more or less large, on the underground parts, such infected plants yield smaller, 'bulbs' which do not keep well.

**Fungus:** During the initial stage giant cells scattered amongst cell of ordinary size. The contents of the giant cells consists of either a frothy granular mass of protoplasm or a large number of tiny spheres the spores. The germination of these spores can be observed under the microscope. A small opening appears in the wall of a spore, through which its protoplasm is extruded. When quite free this speck of protoplasm swims about for a few hours by means of a hair like cilium, it thereafter loses its cilium and becomes a creeping amoebae like organism termed as myxamoebae. In nature rotting of the host tissues sets the spores free in the soil where they germinate. The myxamoebae move through the soil, either swimming in the films of water present in all ordinary moist soils, or creeping through the interstices. Some of the organisms may chance to reach the root of a cruciferous plant, but probably the greater portion of them fail to do so, and perish. The myxamoebae are believed to gain entrance to the host through the root hairs.

When an organism gets inside a root cell it feeds on the contents and this particular cell seems to make a special call on the food supplies of the plant for it gradually increases in size. But the parasite eventually devours all the cell protoplasm and the cell becomes filled with fungus protoplasm – the plasmodium which is the vegetative state of the slime fungus. The plasmodium is able to move from the giant cell into neighbouring cells. Causing these to become giant cells also. The formation of giant cells causes the roots to become swollen, and when the extension of the plasmodium proceeds in a definite direction protuberances are formed. The development of the plasmodium makes a demand on the manufactures food of the plant, and this demand increases until the supply is insufficient for both plant and parasite, when this point is reached. The growth of the plant which has been gradually declining ceases altogether and the plant dies (Fig. 6.2).

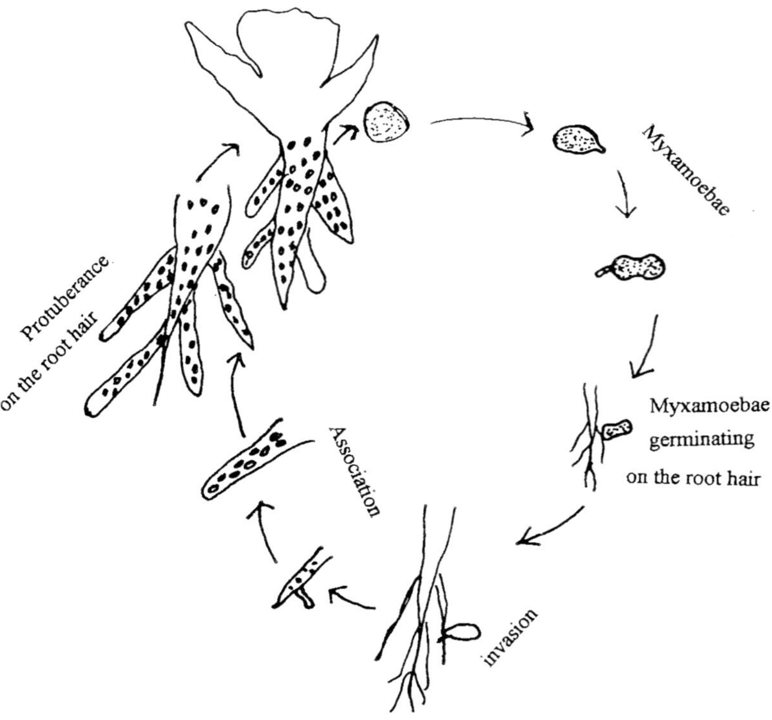


**Fig. 6.2:** Life cycle of *Plasmodiophora brassicae*

As each portion of the plasmodium in an exhausted cell matures it divides into a large number of parts, each of which becomes spherical and surrounds itself with a definite wall, these are the spores, which lie free in the giant cell, and are discharged into the soil. They germinate in the spring following their liberation in the succeeding years.

They germinate in the spring following their liberation, or in the succeeding years, whether the spores remain in the soil without germinating and long they may do so, or whether the myxamoebae can live as saprophytes for a time, are points still undetermined. However this longevity, together with the presence of cruciferous weeds, enables the fungus to continue its existence when cruciferous crops are grown only once in four or five years.

# 7. CHYTRIDIOMYCETES

Chytridiomycetes are more closely related to fungi. Main feature that occurs in chytridiomycetes, and absent in other fungi that remain to be considered is the zoospore, which has a single smooth posterior flagellum. This could be a primitive feature of the fungi, lost in the group that adopted a more terrestrial life style.

About 600 species of chytridiomycetes are known, it is classified into four order. Two of these are Blastocladiales and Chytridiales.

**Blastocladiales:** Members of the order Blastocladiales are mainly saprotrophs living on plant or animal debris in fresh water, mud or soil. Most produce true mycelium, which branches dichotomously and is divided into compartments by occasional partitions, pseudosepta, differing in chemical composition from the hyphal wall. The sexual process consists of the fusion of a pair of uninucleate gametes, both of which are motile. *Allomyces* and *Blastocladiella* are the two genus which are studied in more detail. Another genus *Coelomomyces* is an obligate parasite of arthropods.

**Anaerobic Rumen fungi:** Ruminant animals, such as sheep and cattle are highly effective consumers of plant biomass. This effectiveness is due to the activity of microorganisms in the rumen, a specialized region of the gut conditions in the rumen are totally anaerobic, and the study of rumen microbes require straingent exclusion of oxygen both in the sampling and processing of rumen contents and in the subsequent culture of the rumen organisms. Obligately anaerobic bacteria and protozoa have long been known to be abundant in the rumen. Anaerobic rumen fungi were assigned to the order spizellomycetales. Three genera, *Neocallimastic*, *Piromonas* and *Sphaeromonas* were recognized on the basis of zoospore form and ultrastructure.

**Chytridiales:** In this class fungus body may consist of one cell with a cell wall and with or without nucleate rhizoids which either becomes directly a zoosporangium or gametangium which empties its contents into an external zoosporangium or gametangium. Fungus have true coenocytic mycelium containing tropic hyphae with in the substratum. True septa is absent in it. Fungus is entirely contained in the cell of its host and completely lack rhizoids or haustoria, it is called holocarpic, when it is external to the substratum or internal and gains its nourishment by means of rhizoids or haustoria it is called Eucarpic.

Chytridiales are comparatively simple in their structure. This order includes a member of families of fungi that are either largely aquatic or depend upon the presence of water for their dispersal. They are either parasitic in the roots, stems and leaves of higher plants, in algae, in fungi, in the egg or larvae of worms. Arthropods or simpler animals or perhaps more often saprophytic in dead plant or animal material. They are abundant in some types of soils and in fresh and salt water. Life history of these fungi is simple– posteriorly uniflagellate zoospore after a period of swimming settles down on the substratum and penetrates it or encysts on the outside and then enters its wholly or merely as a more or less extensive haustorium or system of rhizoids. It produces a cell wall of its own. Single nucleus divides repeatedly as the cell enlarges, and eventually the cytoplasm fragments into very many uninucleate zoospores or naked gametes into uninucleate cells which produce cell walls and sooner or later divide internally into zoospores or naked gametes. Sexual reproduction takes place by the union of two cells attached to or reducing with in the substratum or by the union of two cells by means of a rhizoid through which a gamete nucleus passes. Zoospore or motile gametes escape through an exit papilla or tube whose apex softens and permits the motile cells to push out or escape through a sort of cap that opens like a trap door, so called operculum. Operculum is at the apex of papilla or tube, but it may be formed with in tube some distance from the apex, which softens and deliquesces much as in the inoperculate forms.

Harder (1937) reported that in one species of the family Rhizidiaceae different individuals possess cellulose or chitin perhaps in the cells of different ages. Nabel (1939) demonstrated the presence of chitin but not of cellulose in some species of *Rozella* and *Synchytrium*. Schwartz and cook (1928) report the presence of cellulose in the cell wall of *Olpidium radicale* while Scherffel (1925) states that in certain structures of *Micromycopris cristata* and *Synchytrium mercurialis* cellulose is present.

Sparrow (1942) makes the primary division of this order, on the basis of mode of escape of the zoospores, in two series. Inoperculate and operculate. Sparrow recognizes three families, the relationship of the second of which is some what doubtful. They are distinguished as follows.

**Olpidiaceae:** Vegetative cell enlarging to form a single sporangium.

**Achlyogetonaceae:** Vegetative cell elongated and by septation forming a linear series of sporangia.

**Synchytriaceae:** Vegetative cell dividing internally into numerous sporangia or the formed with in and out growth from vegetative cell.

**Family Olpidiaceae:** This family contains six or more genera, totaling forty or more well substantiated species and many more described species whose status according to sparrow is uncertain. They are entirely endobiotic–living entirely with in the host cell or tissues. They are most frequently parasitic but may occur as saprophytes with the dead host cells. Posteriorly uniflagellate zoospores settles on the exterior of the host and the flagellum disappears, in most cases being withdrawn into the body of cell which is then becomes covered with a thin wall. A slender infection tube grows through the host cell wall and the contents of the encysted zoospores passes in the host. The empty cyst soon disappears.

**Genus—*Olpidium:*** *Olpidium brassicae* commonly attacks the roots of cabbage is capable of attacking a wide variety of monocots and dicots. *Olpidium viciae* is a parasite of leaves and stems of *Vicia unijuga*. From the infected parts of the host. Zoospores escape from the zoosporangium through an exit tube. After a period of swarming, they encyst on the surface of the host. Infection takes place via a minute pore digested in the host cell wall through which the protoplast of the parasites enters the host cell, leaving the cyst on the outside.

***Physoderma:*** All members of this genus are obligate plant parasites of vascular plants. *Physoderma maydis* cause, brown spot disease of corn, capable of causing significant losses when corn is grown in warmer regions having abundant moisture. It is disseminated by means of its resistant sporangia. Meiosis occurs during the germination of resistant sporangia that release numerous motile cells or meiospore.

**Order-Monoblepharidales:** This is a small order consisting of only a few species, most of which are saprobes found growing in fresh water on submerged twigs and fruits. Some species can be isolated by baiting soil samples that have been placed in water with hemp or sesamum seeds.

***Monoblepharis polymorpha:*** the somatic thallus consists of well developed branched hyphae, the protoplasm of which is highly vacuolated, so that it appears foamy. This foamy appearance is characteristic of *Monoblepharella* and *Gonapodya*. Elongated zoosporangia typical of those found in many of the water molds. They are generally no larger in diameter than the somatic hyphae. The sporangia are subtended by the septum. Zoospores are released from the tip of the sporangium, swim for a time, become rounded and germinate each by a germ tube, forming a new mycelium.

Same thallus that produces a sporangia produces gametangia when subjected to higher temperatures. Gametangia are easily distinguishable as male and female, with the narrow, elongated male gametangium being borne on the rounded, large female gametangium. A number of uniflagellate gametes are formed with in the male gametangium and later released. Protoplast of female gametangium becomes rounded and enlarged. Male gametes are released from the male gametangium, they swim or creep over the female gametangium. A single male gamete enters the female gametangium through a papilla present in the wall, penetrates the enlarged female cell and fuses with it. After fusion the large protoplast partially emerges from the female gametangium and attacked by a hyaline collar secretes a thick wall around itself, after karyogamy becomes the zygote. The zygote germinates under favourable conditions by producing a hypha. Hypha develops into a thallus. Meiosis probably takes place during the germination of the resting sporangium.

**Synchytriaceae:** Genera belonging to the family are exclusively parasitic possessing a unicellular endophytic and holographic thallus which at maturity gets converted into a sours of reproductive organs. Sporangia are usually formed with in the host cells which in the beginning are surrounded by a common membrane. Zoospores are posteriorly uniflagellate. Sexual reproduction where known, takes place by conjugation of isogamous flagellated gametes resulting in the development of a thick walled resting spore which on germination, function as a sporangium. Three genera are included in this family *i.e. Synchytrium, Endodesmidium* and *Micromycopsis.*

***Synchytrium:*** *Synchytrium* is the largest genus of the family Synchytriaceae, sepresented by nearly 200 species. It has been reported from all the parts of the world both tropical and temperate. They are chiefly parasitic on flowering plants causing development of galls of varying sizes on leaves, petioles, stems, buds and flowers sometimes causing disease of considerable economic significance. Some species are parasitic on mosses and ferns also. Host plants in wet situation, are more easily liable to attack. Some aquatic species of *Synchytrium* also parasitise algae (*S. spirogyrae* and *S. ovalis*).

***Synchytrium endobioticum***

It is an obligate parasite on the black wart of potato. Infection with the parasite produces large or small greenish yellow warty outgrowths on the tubers, hence the name wart disease. The aerial stem and leaves may also be sometimes infected but the galls produced on these are relatively smaller. In advanced stages of the disease. The wart and on tubers rot away exuding a putrid smelling dark liquid.

The zoospores which are released in great numbers from infected tissues of the host, swim actively in the film of water present in the soil. In presence of the host these naked uniflagellate level, dissolve the wall making a minute pore through which it enters the host cell. During this process, it looses its flagellum and can be recognized as an amoeboid structure with in the host cell. The presence of the parasite stimulates the cell to hypertrophic growth. (Fig. 7.1 A-E).

**Vegetative structure:** The parasite absorb food from the surrounding protoplasm and enlarges considerably. Its nucleus also becomes greatly enlarged and prominent. Having attained a certain size, a two layered thick wall is secreted round the thallus. This walled structure is also sometimes known as prosorus (Fig. 7.1 F). As a result of the hypertrophy the infected cell is greatly enlarged. Meanwhile the surrounding epidermal and cortical cells are also stimulated to division and hypertrophic growth resulting in the formation of the typical warty out growths.

**Asexual reproduction:** The uninucleate, thick walled prosorus remains in the infected cell for sometime after which its wall ruptures and contents emerge out with in a thin hyaline membrane furnished by the expanding inner layer. The nucleus now starts dividing mitotically and when about 32 nuclei have been formed. Cleavaging of the protoplast takes place dividing, the prosorus contents into 4–5 or multinucleate segments walls are laid down round the cleavage zones and each of these segments develop into a sporangium. Thus a sporangial sorus develops from each prosorus. Repeated nuclear divisions occur in the sporangia until 200–300 nuclei have been formed. Cleavaging takes place and sporangial contents get segmented into as many units as there are nuclei. Each unit with this nucleus and the surrounding cytoplasm, develops into a uniflagellate zoospore (Fig. 7.1 G-L). When mature, the sporangia absorb water and swell up and are forced out of the sorus after rupture of the host wall. The zoospores are liberated through one or more rudimentary papillae or slits that develop in the sporangial wall (Fig. 7.1. M-P).

Zoospores swim actively in the surrounding water and on coming in contact with the host surface, withdraw the flagellum and enter the epidermal cells of the host. Whether the sorus will behave as a zoosporangial sorus or a gametangial one depends on the environmental conditions. A sporangial sorus develops if water is present abundantly and in the lack of the water the same sorus behaves as gametangial. Zoospores that are retarded in emergence may also behave as gametes.

**Sexual Reproduction:** Zoospores from over ripe sporangia, or from sporangia developed during periods of relative drought or those retarded in emergence, may also behave as planogametes. These gametes are uniflagellate like the zoospores but are some what smaller in size. They swim for sometime, form pairs and fuse to develop a biflagellate zygote. Nuclear fusion follows plasmogamy. There are evidences to show that gametes from the same gametangium donot fuse although, those from different gametangia of the some sorus may.

The biflagellate zygotes thus formed, swim for sometime and then come to rest on the surface of the host whose epidermal cell is penetrated in the same way as the zoospores. The zygote settles down at the bottom of the cell. The infected cell and its neighbouring cells are stimulated to hyperplastic division and with the developing pressure the zygote is buried deeper in the tissue. It enlarges with in the infected cell and secretes a two layered thick wall around it. This double walled structure is known as resting spore, which on germination, functions as a sporangium hence also termed as resting period of rest when the diploid nucleus under goes successive divisions giving rise to a large number of nuclei. The first of these divisions in believed to be meiotic. Cleavaging occur in the protoplasm of the sporangium resulting in the formation of many zoospores as there are nuclei. These zoospores are also posteriorly uniflagellate structure but are slightly larger in size than the zoospores produced from the sporangial sorus. They function in the same manner as the asexual zoospores (Fig. 7.1 N-V).

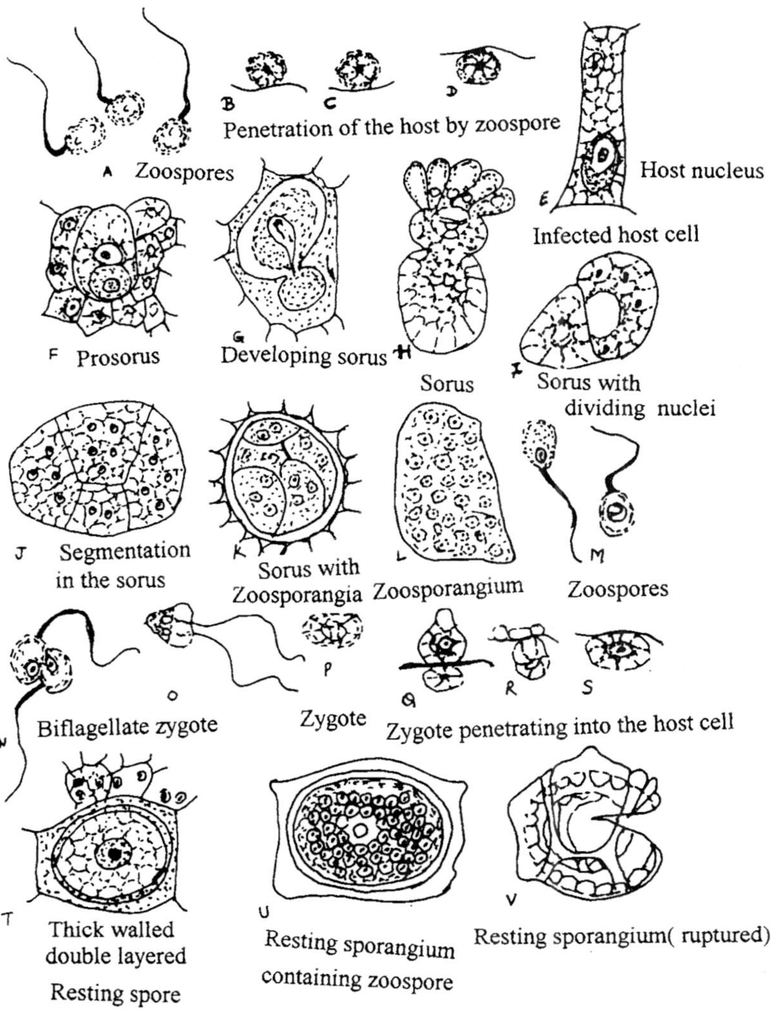


**Fig. 7.1:** Stages in the life history of *Synchytrium endobioticum*

**Family Catenariaceae:** This family is parasitic in worms of fungi or saprophytic in various other plant or animal subtracts. The plant body at first is tubular, mostly unbranched, coenocytic swelling at more or less regular intervals to form reproductive organs which are connected by short, narrow thallus. Septate at each end and sometimes in the middle. These reproductive bodies are either thin walled zoosporangia discharging by exit tubes or on more exhausted media, thick walled resting spores usually free from the hyphal membrane. These resting sporangia are smooth or minutely granular. During germination, outer wall cracks open and a tube emerges through which the zoospores are discharged. Rhizoids are produced at almost any point on the thallus. Sexual reproduction occurs by the union of equal, motile gametes. One genus only is known *Catenaria.* Two well studied species are *C. anguillulae* and *C. allomycis*.

*Catenaria anguillulae* is polycentric and eucarpic producing a thallus consisting of branched or unbranched septate hyphae with rhizoids. Zoosporangia and resting spores develop from swellings that arise at more or less regular intervals along the hyphae. Other species of Catenaria have been reported as parasites in the eggs and egg masses of midges.

**Family Blastocladiaceae:** Vegetative portion of the plants of the family consists of a more or less extensive system of tapering and branching rhizoids and of a globular or clavate external portion from which may arise directly the sporangia and gametangia or a system of branches on which these are borne.

*Blastocladiella:* In *Blastocladiella stubenii* the plant body of the sporophyte is spherical, with tapering, much branched rhizoids emerging at all sides. It bears a thin walled zoosporangium given with one to several discharge tubes or a dark, thick walled, resting sporangium which produces swarm cells that give rise to indistinguishable male or female gamete similar to but smaller than the sporophytes. Emerging gametes are equal in size and cannot be distinguished by colour. They fuse in pairs and at once produce the sporophytes.

# 8. OOMYCETES

Oomycetes are a common group of organisms found all over the world in both fresh and salt waters as well in Terrestrial environments. Commonly they are known as "water molds"--grow in fresh water, primarily in rivers, ponds and lakes. They can also be found in shallow water, stagnant water and at the bank of rivers. Most acquatic forms are saprobic, found on the dead plants and animals, play a great role in the degradation and recycling of nutrients in acquatic ecosystems.

The characteristic feature of oomycetes is the typically oogamous type of sexual reproduction in which the female gametangium, termed as oogonium is a globose structure containing one or more oosphere (egg) and these are fertilised by the male gametangium antheridium that reaches the oosphere through a fertilisation tube. The resultant zygote which originates and matures with in the oogonial wall, is termed as oospore. This sexual reproduction is known as heterogamous type.

The oosphere then develops into the oospore characteristic of this class. Each oospore has a single diploid nucleus. When oospore germinates it gives rise to a mycelium that is diploid, in contrast to the haploid mycelium of most other fungi. Another characteristic of the oomycetes that distinguishes them from other Eumycota is the biflagellate zoospore this has a forwardly directed tinsel flagellum—a flagellum with a row of hairs on each side when viewed with the electron microscope—as well as a posteriorly directed smooth flagellum. Their cell walls are mainly consist of glucans and cellulose. In some species chitin is absent.

Asexual reproduction takes place by means of biflagellate zoospores with a longer tinsel flagellum directed forward and a shorter whiplash flagellum directed backward. Sporangia may behave as conidia and germinate directly by a germ tube which give rise to the mycelium.

The somatic structure of the fungi range from a unicellular thallus to a profusely branched copious, filamentous mycelium which grow abundantly in the substratum. Mycelium is coenocytic and a profusely branched filamentous structure mostly eucarpic. Hyphal walls generally give reaction for cellulose. The hyphae of the aquatic forms sometimes form extensive growths on their natural substrates that may be large enough to be seen with the naked eye. The oomycetes with the simplest structure are acquatic fungi either free living or parasitic on algae, water molds, small animals and other forms of acquatic life. The most complex of the oomycetes are terrestrial parasites of plants passing their entire life history in the host and depending on the wind-for dissemination of their spores or spore like sporangia. Hyphae of species that attack higher plants either in grow an intracellular or an intercellular fashion. Hyphae of facultative plant parasites tend to grow indiscriminately into and through the head and dying cells of their hosts. Obligate parasites tend to remain between hosts cells, giving rise to specialized hyphal branches termed haustoria that extend through the host cell wall and invaginate the host cell plasma membrane. Plant parasitic oomycetes produce either peg like spherical or lobed haustoria.

Oomycetes are classified into five orders *i.e.* Leptomitales, Saprolegniales, Rhipidiales, Lagenidiales and Peronosporales.

**Order Saprolegniales:** Most of the members live in water. They are confined to fresh, clear water, some species are able to with stand a certain degree of salinity and live in the brackish waters of estuaries–*Saprolegina, Achlya* and *Aphanomyces.*

**Order Lagenidiales:** This is a small group of primarily aquatic fungi parasitic on other aquatic or semi aquatic life forms including algae, watermolds and small animals such as rotifers, nematodes and arthropods. Thalli of these organisms are endobiotic and monocentric consisting of either single cells or very short, unbranched or sparingly branched tubular filaments. Sexual reproduction takes place by gametangial copulation, with or without a fertilization tube and results in the formation of a thick walled oospore. Example–*Olpidiopsis* and *Lagenidium.*

**Order Rhipidiales:** These fungi produce inflated and elongated monocentric thalli with rhizoids attaching the thallus to the substrate. Leptomitales occur on plant materials submerged in clear, unpolluted water species of Rhipidiales. *Aqualinderella fermentans* grow in stagnant and on polluted waters having low levels of oxygen. Other genera are *Sapromyces, Asaispora, Mindeniella* and *Rhipidium.*

**Order Perenosporales:** This large order include aquatic, amphibious and terrestrial species, culminating in a group of highly specialized obligate parasites. Damping off fungi, the white rusts and downy mildews all belong to this order. Example *Perenospora, Phytophthora, Plasmopara, Pythium, Bremia* and *Albugo* (Fig. 8.1. A–C).

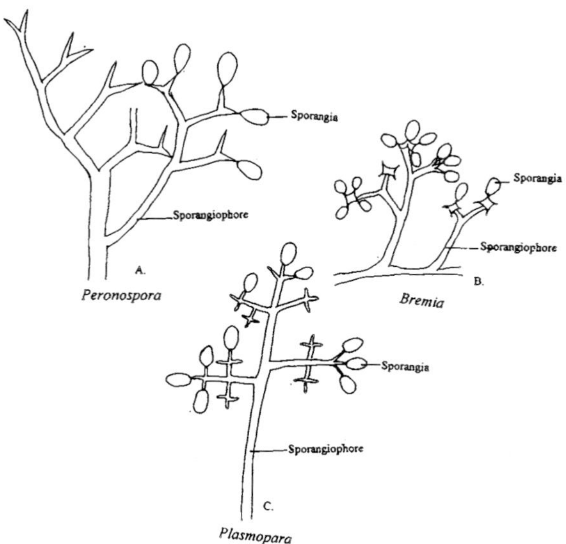


**Fig. 8.1:** Sporangiophore and Sporangia of A. *Perenospora* B. *Bremia* C. *Plasmopara*

In *Plasmopara* branches of sporangiophores and their subdivisions occur typically at right angles and are irregularly spaced. In case of *Bremia* Sporangiophores are dichotomously branched like *Perenospora* but the tips of the branches are expanded into cup shaped apophyses with four sterigmata, each bearing the sporangia along their margins.

**Order Leptomitales:** This order contains only a few saprobic, aquatic species. Cellular granules are present in the hyphae and often plug the constrictions, giving the hyphae a septate appearance.

**Albuginaceae**

Members of this family are obligate parasites on herbaceous angiosperms possessing intercellular mycelium with globular haustoria. There are several species of *Albugo*, the only genus in this family. *Albugo candida,* which attacks crucifers and capparidaceae and also on leaves of *Reseda alba.* This is the only one causing disease attaining economically significant proportions. Some of the other species commonly found *A. bliti* on Amaranthaceae, *A. occidentalis* on spinach, *A. ipomoeae panduranae* on *Impomoea, A. tragopogonis* on compositae plants. Distribution is world wide where host are present.

**Life History of *Albugo*:** Disease caused by this fungus is commonly known as white rust on account of the shiny white pustules that are formed on the affected leaves mainly on their lower surface. Sometimes infection also produces deformations of the flowering shoots, floral parts and even the fruits. The attached parts often show marked hypertrophy, especially of the inflorescence. In rare instances galls have been found on roots of radish, containing oospores and globular haustoria (Fig. 8.2).

**Vegetative Structure:** Fungus is endoparasitic and the mycelium, strictly intercellular, is composed of well branched, hyphae is aseptate, coenocytic in nature, small globose to knob like haustoria one to several in each host cell. Large number of nuclei are present in the cytoplasm, with oil drops and reserve food in the form of glycogen bodies.

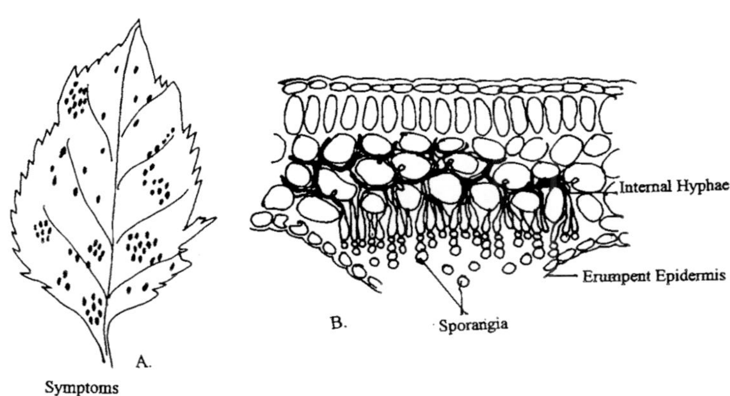


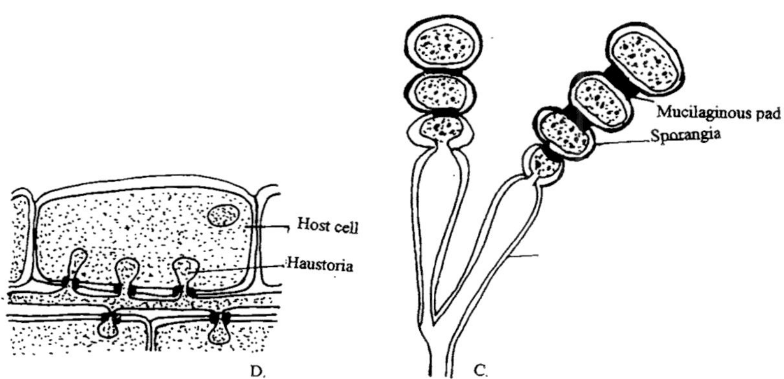


**Fig. 8.2:** A. Symptoms of *Albugo candida* of leaf of *Brassica campestris*. B. V.S. of leaf showing infection. C. Detailed Structure Sporangiophore bearing Sporangia. D. Entry of haustoria into the host cell

**Asexual Reproduction:** Asexual reproductive organs are sporangia, that develop beneath the epidermis of the host on club shaped sporangiophores. Rapidly branching mycelium enters into the intercellular spaces of the host tissue, after a certain stage of maturity, it develops into the sporangiophores. These are produced by hyphal branches that emerge out of the host tissue at various places beneath the epidermis and are arranged in parallel fashion, palisade like cells are present. Each sporangiophore gives rise to several sporangia that it produces in success one below the other, so that a chain of sporangia is formed with the oldest at the tip of the chain and the youngest at the base. When sporangia becomes mature, they accumulate in the space between the sporangiophore and epidermis (Fig. 8.3).

By the growth of mycelium and sporangia pressure is exerted by them on the epidermis, by which the epidermis bulge out and the epidermis rupture. As the epidermis ruptures, sporangia comes out on the surface and forms a layer of white crust. Mucilaginous pad is present in between the sporangia. Sporangia are round, thin walled and multinucleate. Dispersal of sporangia takes place by means of wind and water. When the temperature becomes favourable, 4–12 zoospores comes out from the sporangia. Germination of zoospores takes place by the formation of germ tubes. Zoospores are kidney shaped or slightly concavo – convex, have a vacoule and the two laterally inserted flagella. They swim actively for sometime then come to rest, recreate a wall and undergo a period of encystment. On germination they put forth a short germ tube, which enters the epidermis through stomata and produces a well developed intercellular mycelium (Fig. 8.3 F-J).

**Sexual Reproduction:** Mycelium which bears sporangia can bear sex organ also. Sexual reproduction is oogamous type and takes place by forming antheridia and oogonia (Fig. 8.4). Sex organs are in the deeper tissues which are swollen and malformed. Sexual reproduction in its gross aspects is similar in all aspects, but the cytological details of fertilization seem to fall into at least three patterns (number of nuclei takes part in fertilisation). At the time of fertilisation functional nuclei can be one or can be in groups.

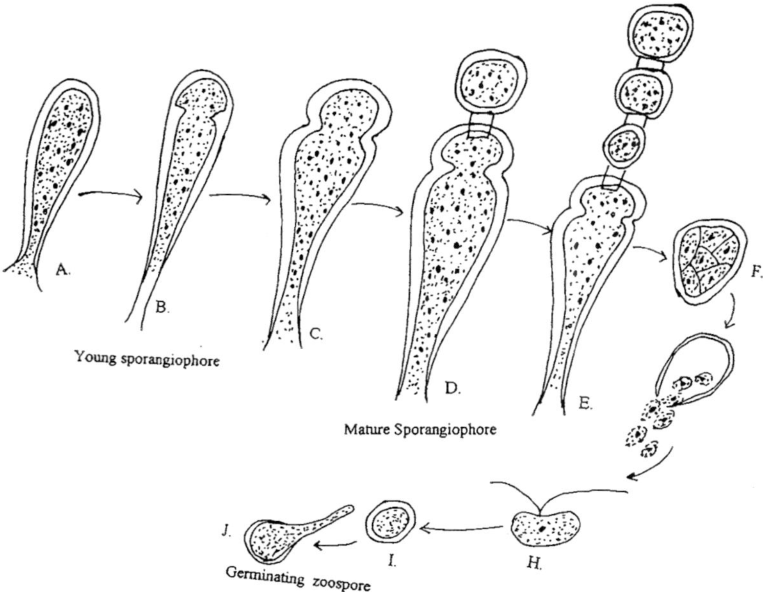


**Fig. 8.3:** Early stages in Development of sporangiophore and sporangia (Asexual Reproduction)

Oogonia and antheridia develop into deeper tissues of host. Both the sex organs develop terminally over hyphal branches and initially they were multinucleate. Oogonia is globular is structure and contain about 100 nuclei. Antheridia is clavate or curved is structure and it is attached laterally to the oogomia, and have 6–12 nuclei. Nuclei are distributed uniformly in the cytoplasm and under go mitotic divison. After nuclear divison, oogonium get organised into two layer. Peripheral layer is known as periplasm and dense cytoplasm is known as ooplasm or the egg. In the center, one nuclei remains and rest of the nuclei migrated to the periphery. Nuclei of the periplasm under go mitotic divison. Divison takes place is such a fashion that one nuclei falls in the periplasm and other in the ooplasm region. When the egg mature and ripe for fertilisation all the nuclei in the peripheral region degenerate, in the centre one large nucleus is left. If the antheridium also simultaneous divison and degeneration of nuclei takes place and only one remain functional. After this the receptive papilla develops in the antheridium at the point of contact. By piercing the periplasm receptive papilla makes its way into the egg. Functional male nucleus migrates to the tip of fertilisation tube, as tips ruptures, male nucleus discharges into the egg where it immediately fuses with the egg nucleus. Fertilisation tube and antheridia remain attached to the egg. Thick wall is formed around the egg and become seprated into endospore and exospore In the *A. candida* oospore globose, chocolate brown, 30–55μ, epispore thick, verrucose to tuberculate, with low blunt ridges, which are irregularly branched, sometimes it is smooth (Fig. 8.4 D-J).

While in *A. ipomoeae panduranae* oospores caulicolous, spherical, light yellow brown 20–40μ epispore papillate or with irregular more or less curved ridges.

In *A. tragopogonis* oospores are spherical dark brown to almost black at maturity, 44–57μ, epispore reticulate, densely covered by low tubercuiate or spinulose warts, meshes 2–3μ across.

In case of *A. bliti, A. portulucae, A. platensis* many functional nuclei are present in the sex organs. Functional nuclei may be 100–300 in number. So that compound oospore is formed. Zygote

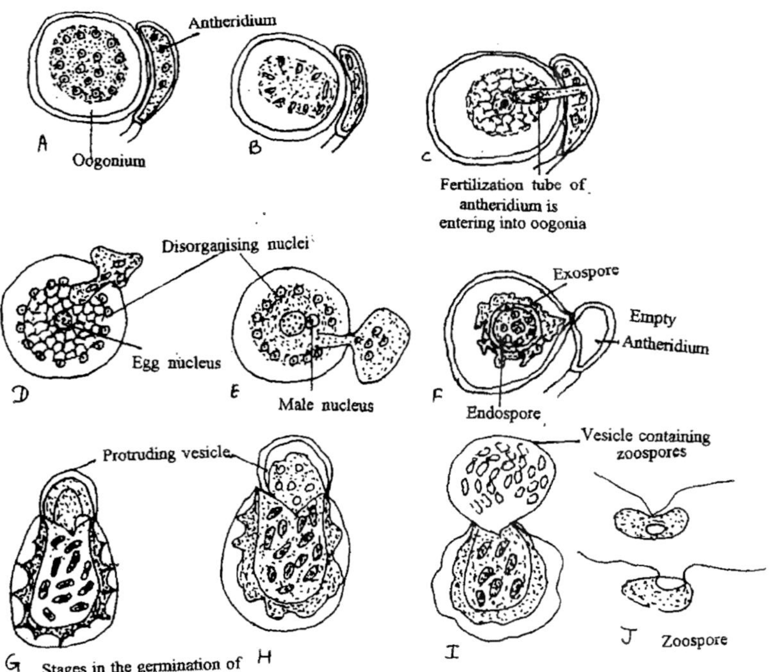


Stages in the germination of Oospore by rupturing of the Exospore

nucleus undergoes several divison and the oospore rests as a multinucleate structure. After a long resting period and on the return of favourable weather conditions, the oospore germinate. Nuclei may under go one or several mitotic divison and the protoplast of the oospore divides into a large number of uninucleate portions, each giving rise to zoospore. Epispore ruptures, a thin vesicle protrude out of the endospore and the mass of zoospores in the surrounding fluid. The biflagellate zoospores are similar in shape and size to those produced in the sporangium. They swim actively for sometime, encyst and finally germinate by a germ tube to infect the most plant in the usual manner.

***Perenospora:*** It occurs on the many crucifers, *Vicia faba, Lathyrus sativus, Pisum sativum, P. arvense, Allium cepa, A. fistulosum, A. porrum, A. sativum* etc. Commonly it is known as **downy mildew** mycelium is well developed, aseptate, coenocytic mycelium lives with in the tissue of the host plant. They draw nourishment from the host cells by means of haustoria which are large clavate or finger shaped and branched. The haustoria often fill the cell cavity and absorb nourishment from the host cell and pass it on to the growing aseptate coenocytic fungus mycelium.

**Symptoms:** On leaves, elongated faintly yellowish lesions on the inner and outer side of leaves, covered by greyish violet down, leaf tips shrivel, drying up of leaves, beginning with the older and outer leaves and progressing to young leaves until the entire plant may be killed. On seed stalks, similar lesions, especially on the upper part, often causing twisted or topsided stalk growth–Death of leaf and stalk tissue frequently nastened by secondary infection. Common species of *Perenospora* is *P. parasitica, P. destructor* and *P. pisi* (Fig. 8.5 A-B).

**Asexual Reproduction:** Asexual reproduction takes place by means of Sporangiophores. Sporangiophores emerge in clustures of 5–7 from each stoma, long, stiff, straight unbranched for 2/3 or more of their height, 160–750 × 8–13μ 5–8 times dichotomously branched, primary branches straight to slightly curved. Upper ones curved and spreading, ultimate branchlets diverging at obtuse or right angles, pointed unequal, short, 18–22 × 2–3μ, bearing a single sporangium at each end. Sporangia oval to elliptical, narrowed a little towards attached end, 15–30 × 15–20μ, pale violet to pale greyish in mass at maturity fresh hyaline. The sporangiophores species have a determinate growth. Sporangiophore reaches maturity, stops growing and then produces a crop of sporangia on sterigmata at the apices of its branches. All the sporangia are therefore approximately of the same age. These sporangia germinate by germ tubes and have been called conidia by some researchers (Fig. 8.5 C).

**Sexual Reproduction:** It is oogamous type and takes place toward the end of the growing season when the weather is dry and the environmental conditions are adverse. Mycelium of conidia penetrates deeper into the host tissue and produces the sex organs. They are commonly found in the pith on the hyphae in the larger intercellular spaces especially of the stem. Both are multinucleate at first and arise as terminal swellings at the hyphal tips of closely adjoining hyphae (Fig. 8.5 D).

**Oogonium:** The oogonium is more or less globose structure. It is seprated from the supporting hypha by a cross wall. When young it has a dense multinucleate protoplast. Nuclei are evenly distributed. Oogonium wall differentiate into two layer, outer vacuolated layer is known as periplasm, central dense, less spherical is know as ooplasm. Nuclei in the central part moves towards the outer side, except one functional nuclei which remains in the center. The uninucleate ooplasm functions as an oosphere or egg. Only a single egg is formed in the oogonium.

**Antheridium:** Antheridium is paragynous. It is tubular in form and is multinucleate at first. It is separated from the supporting hypha by a cross wall. At maturity all nuclei except one degenrate.

**Fertilization:** As antheridium comes in contact of oogonium, fertilisation tube develops at the point of contact. Fertilisation tube pierces through the oogonial wall and the periplasm. Finally its tips dissolves discharging the male nucleus and a small amount of cytoplasm into the oosphere. The male and female nuclei fuse. Zygote thus formed, have a thick wall. It is globose and yellowish brown in colour. Many layered oospore enters upon a period of rest extending over several months.

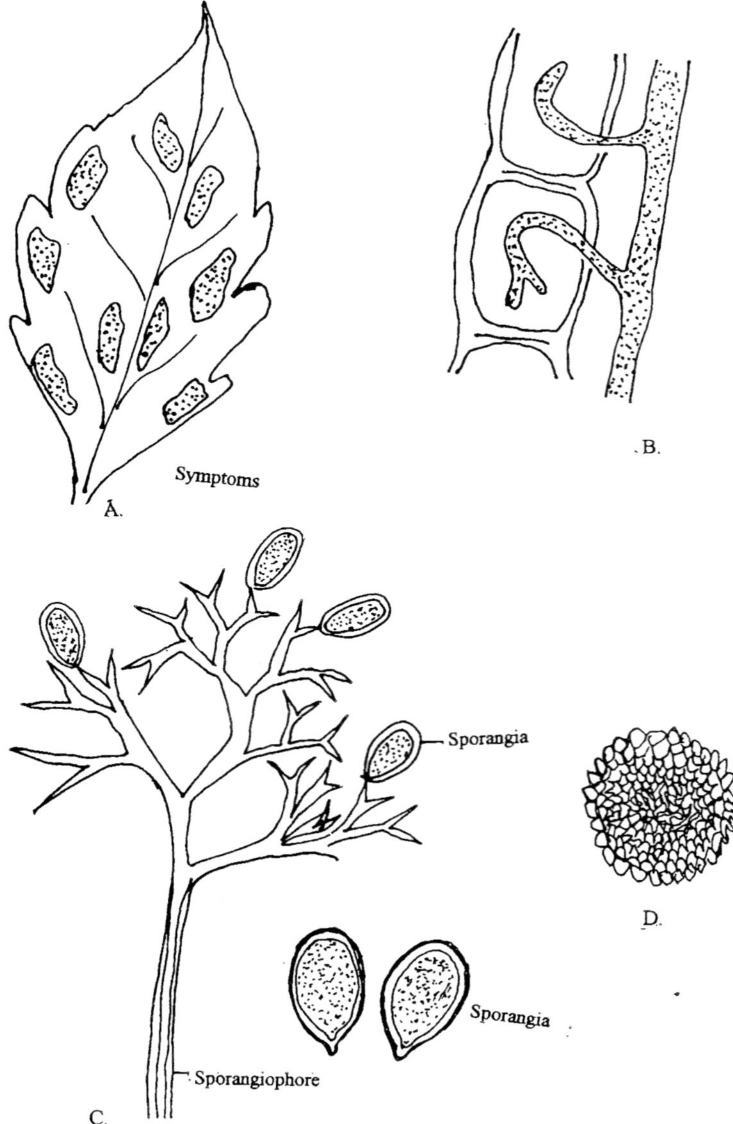


**Fig. 8.5:** A. Symptoms of *Perenospora* on the host. B. Mycelium entring in the epidermal cell. C. Sporangiophore bearing sporaniga. D. Oospore

After a period of rest and with return of conditions favourable for growth the oospore germinates. The thick oospore wall bursts and the protoplast emerges in the form of a germ tube which brings about fresh infection. The first two nuclear divisons at the time of oospore germination constitute meiosis.

***Phytophthora:*** This species cause a wide variety of diseases on a tremendous number of hosts including both herbaceous and woody species.

*Phytophthora infestans* is a notorious fungi causing a late blight of potato. This disease contributed to the great famine in Ireland in 1845 and 1846, resulted deaths of perhaps 1 million Irish people and emigration of upto 1.5 billion individuals, to North America. This genera is represented by 20 species.

**Habit and Occurrence:** *Phytophthora* are parasitic on higher plants mostly angiosperms and cause diseases of economic significance. Some species commonly live as saprophytes in the soil, but in presence of a suitable host may develop as parasite. *Phytophthora infestans* causes **late blight of potato**, *P. colocasiae* on *colocasia esculenta, P. faberi* on (*Hevea brasiliensis*) cocoa and coconut plants, *P. arecae* on *Areca catechu*, *P. palmivora* causing bud rot of toddy and coconut palms and *P. parasitica* on castor and cotton plants etc.

**Symptoms:** The disease manifests after the flowering period usually in the month of January. Patches are small initially, later black patches appear at the margin and tips of the leaflets. Patches gradually enlarges and spread over the entire surface of the leaf lets. Soon the disease invades the petioles and stems finally killing the entire tops of plants which fall over in a rotten pulp. Excess of water coupled with a range of temperature between 22–23°C promotes optimum fungal growth (Fig. 8.6).

**Vegetative Structure:** Mycelium consists of hyaline, irregularly branched and aseptate hyphae, septa often developing in older parts or to delimit the reproductive structures. The hyphae, within the host are mostly intracellular directly penetrating and killing the invaded cells. In some species they may be intercellular also with haustoria piercing the adjacent cells for nourishment.

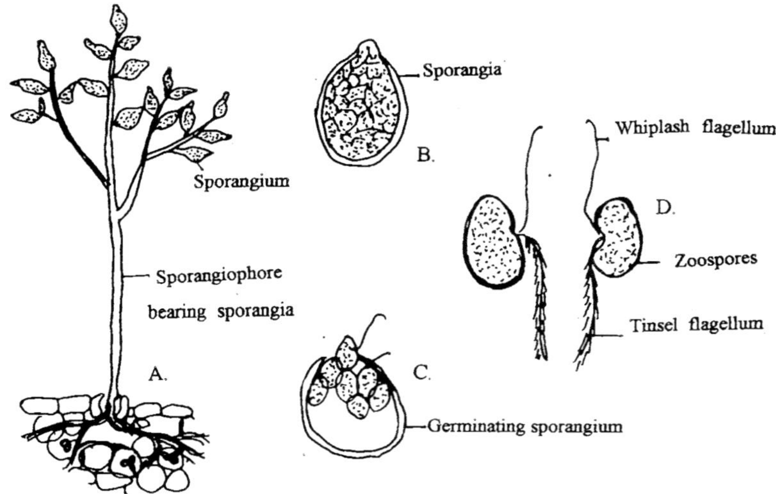


**Fig. 8.6:** Asexual reproduction in *Phytophthora*

In *P. infestans,* perenating mycelium has also been observed, the hyphae lie dormant in the infected tubers for certain periods (Fig. 8.7 E). When such tubers are sown in the soil the mycelium again becomes active. It grows alongwith the sprouting shoot infecting them and sporulating in the leaves.

**Asexual Reproduction:** Asexual reproduction takes place by means of zoospores. They are formed inside the zoosporangia. Zoosporangia are borne on aerial sporangiophores, which in parasitic forms. Commonly emerge through the stomata, but sometimes they pierce the epidermal cells. Sporangiophores can occur singly or in groups. They may be simple and unbranched bearing a single terminal zoosporanigum as in *P. colocasiae,* may be branched sympodially bearing a number of zoosporangia at tips of terminal or lateral branches as in *P. infestans.* Terminally produced zoosporangia ultimately occupy a lateral position due to the faster apical growth of the sporangiophore branches. All the zoosporangia do not mature simultaneously as the sporangiophore keeps on growing and branching, even after the sporangial development with the result that zoosporangia of different ages exist together on the same sporangiophore (Fig. 8.7 A-D).

Zoosporangia in general are oval or more or less lemon shaped structures bearing terminal papilla. These develop as apical outgrowths of the sporangiophore into which migrates cytoplasm and a number of nuclei. They mature into the characteristic shape, get detached and are usually disseminated through the wind. In several species, there occurs a slight thickening of sporangiophore branch just below the sporangium giving its a characteristic structure *i.e.* swollen and bent.

Sporangial production is greatly influenced by temperature and relative humidity. A temperature between 18–20°C and humidity 100% is supposed to be optimal for their development. At lower humidity they soon lose their viability.

Germination of the zoosporangia takes place in presence of water by putting forth a germ tube when they behave as conidia but more often they produce zoospores. The optimum temperature for germination of the zoosporangium of most species falls between 12–15°C with 2°C to 24°C as the minimum and maximum range. At higher temperatures, they germinate mostly by germ tubes, while lower temperatures favour production of zoospores. Cleavaging of the contents of the zoosporangium takes place and the protoplasm is divided into a number of polyhedral, uninucleate masses which later round up and form the zoospores. No nuclear divison occur with in the zoosporangium and many zoospores are formed as there are nuclei. The sporangial wall ruptures at the papilla and the zoospores escape one by one. In some species, the zoospores pass into a thin walled vesicle which soon ruptures, releasing them in the surrounding water. The zoospores are nearly reniform with two laterally inserted flagella. They swim actively in the film of water for varying length of time, the active period depending somewhat on temperature. After the motile period they eventually come to rest, encyst and germinate by a germ tube which may develop an appresorium to penetrate the host tissue through stomata or epidermal cells. The host thus is infected many times during the asexual life cycle of the parasite.

Sometimes, sporangia behave as conidia and instead of producing zoospores, they develop short conidiophores over which arise one or more conidia.

**Chlamydospores:** They are of rare occurrence but sometimes develop as smooth, spherical, thick walled structures either as terminal or intercalary spores. These have been termed differently by different authors viz., resting conidia or parathenogenetic oospores. However they behave as chlamydospores and on germination give rise to germ tube which develops the new mycelium.

**Sexual Reproduction:** Sexual reproduction is typically oogamous type but depending upon the relative position of antheridia and oogonia. *Phytophthora* can be classified into two distinct groups (*i*) Amphigynous oogonia with basal antheridium attached as a collar. (*ii*) Paragynous–oogonia with antheridia attached laterally. Amphigynous type of antheridia are characteristic of the *P. infestans* group and are found in *P. erythroseptica, P. phaseoli, P. parasitica, P. arecae, P. himalayensis* and others. The paragynous type of antheridia occur in *P. cactorum* and *P. nicotianae.*

Antheridium initial arises as a lateral outgrowth from a hypha. The oogonium initially arise below the antheridium (may be on seprate hyphae). The young oogonium pierces the developing antheridium from below, grows entirely through it and emerges out on the upper side to form a globose structure. Contents from the supporting hypha continuously flow into the oogonium which enlarges to develop as a fully mature spherical oogonium having dense cytoplasm and several nuclei. The antheridium also a multinucleate structure but with fewer nuclei, remains at the base of the oogonium forming a funnel shaped collar. Later, the connecting passage between the oogonium and the underlying hyphae is plugged by a refractive substance. Thus separating the oogonium from the rest of the hyphae. The oogonial protoplasm gradually gets differentiated into the denser central ooplasm and the vacuolated peripheral portion the periplasm. Meanwhile the nuclei of both oogonium and antheridium undergo a mitotic divison. Only one nucleus remains in the ooplasm while the rest migrate towards periplasm where they degenerate. Simultaneously, the nuclei in the antheridium also degenerate leaving behind a single functional male nucleus. A receptive papilla develops in the oogonial wall through which the antheridium puts forth a fertilisation tube. It enters the oogonium and discharges the male nucleus into the ooplasm where after sometime fusion occurs between the male and female nuclei. The fertilised oospore recreates a thick wall and undergoes a period of rest (Fig. 8.7 A-G).

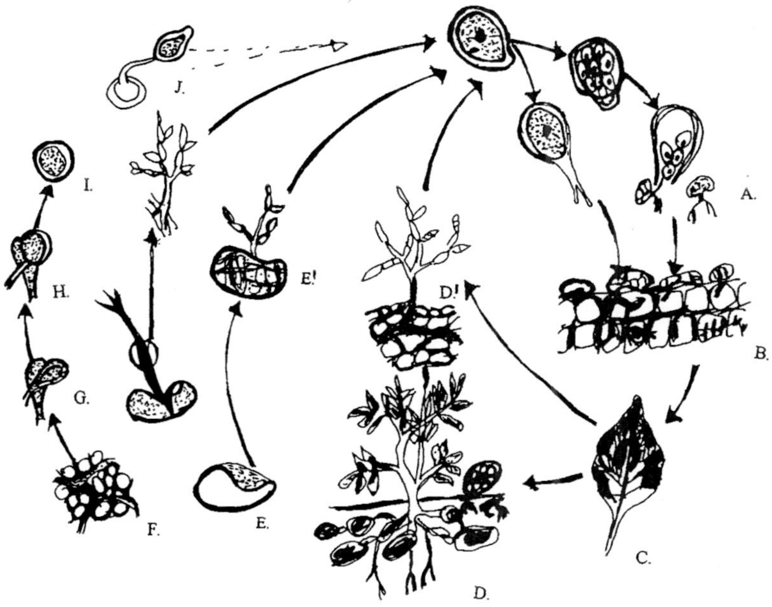


**Fig. 8.7:** A. Escaping zoospores B. Germinating zoospore entring in the leaf through epidermal cells. C. Symptoms on the leaf. D. Tuber and leaf part are infected. D. Tuber and leaf part are infected D. Sporangiophore bearing zoosporangia. E. Infected potato. F-J. sexual reproduction.

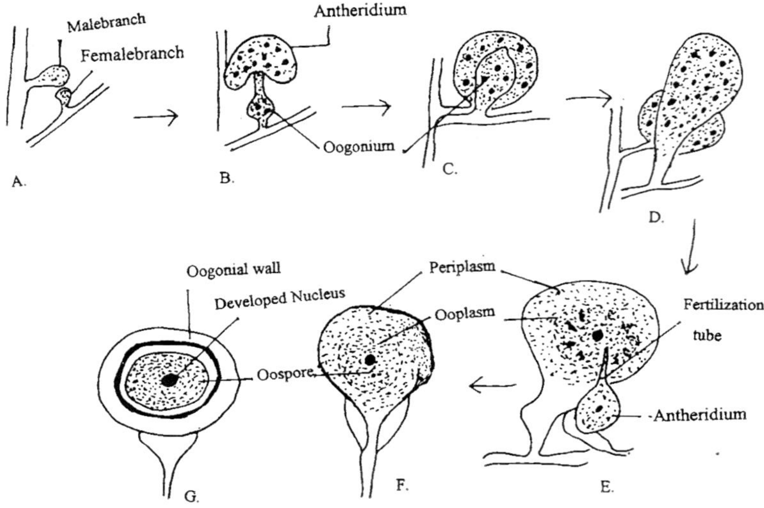


**Fig. 8.7'**: Detailed of Amphigynous of Sexual Reproduction in *P. infestans*. (Detailed of the Life Cycle Fig.8.7 F-J)

In *P.* infestans, another member of the amphigynous group, the oospores rarely develop in nature. The species is known to be heterothallic and the lack of oospores has been explained to be due to the absence of the two mating strains of hyphae commonly occurring in nature. These however have readily been obtained in pure cultures. Oospores of *P. infestans* are also known to develop parthenogenetically in absence of antheridia.

In members of the paragynous group represented by *P. cactorum* the sex organ develop in the usual manner common for most of the oomycetes *i.e.* antheridium occurring laterally to one side of the oogonium. At maturity, antheridium pushes a fertilization tube into the oogonium through which the single male nucleus reaches the egg nucleus and fuses with it after sometime. Oospore is surrounded by the thick wall in the usual way. In *P. cactorum* where the antheridium is typically of paragynous type rare and occasional presence of amphigynous antheridia has also been reported.

The germination of oospores takes place after a long resting period when the fusion nucleus divides successively resulting in formation of large number of nuclei, the first divison being meiotic. A germ tube is formed which usually terminates into a sporangium with zoospores or sometimes may directly develop into a new mycelium.

Spring infection of potato plant originates in diseased potato tubers in which the mycelium survives. The fungus grows into the new tissues sprouting from the potatoes and sporulated on the aerial parts of the plants. Subsequent infection of potato plants takes place by means of sporangia that are transported by water or blown by the wind. Sporangia are extremely susceptible to dessication. When the relative humidity drops much below 100 percent, the sporangia die in a few hours. In the directly by a germ tube that enters through a stoma and infects the leaf, or by means of zoospores. Sporangia are capable of germinating with in a wide range of temperature from 15 to 24°C. However above 20°C the sporangia lose their viability in 1 to 3 hours in dry air and 5 to 15 hours in moist air.

Other factors besides temperature and moisture such as age of sporangia, also affect germination. The method of germination is largely governed by temperature. Low temperatures favour zoospore production, higher temperatures, germ tube production. The optimum temperature for direct germination is 12°C. Zoospores swim in the film of water for 15 minutes at high temperature and upto 24 hours as the temperature decreases.

After the zoospores come to rest they encyst and germinate each by a germ tube. The germ tube produces an appresorium a flattened hyphal, pressing organ from which a minute infection peg grows and enters the epidermal cell of the host. If conditions favourable to penetration of the host last for 10 hours. After penetrating into leaf, the germ tube develops into a profusely branched mycelium that is intercelluar, sending long, curled haustoria into the leaf cells. A few days after infection, if the weather is favorable, numerous sporangiophores emerge from the stomata of the potato leaves and give rise to large numbers of sporangia. These are spread by the wind and infect new plants. A large number of asexual generations are thus produced in one growing season if conditions favour the development of the fungus.

***Pythium:*** This fungus occurs all over the world in the soil. But some species are in acquatic and belongs to the family Phythiaceae. *Pythium* species are the important causes of pre and post emergence damping off diseases of seedlings. This is also known as root rot, fruit rot also. They are parasitic on the seed and seedlings of higher plants. While others are parasitic on algae and as saprophytic on the debris. *Pythium* also causes serious problem to grasses, cabbage, beans and potatoes. *Pythium debaryanum* is one of the most well known species of the genus and is very destructive to young seedlings both of herbaceous and woody plants causing the disease "damping off". Ground level tissue becomes soft and affected the seedlings to bend and tople down and entire shoot decomposes. Plants growing in nurseries and glass house are generally affected by this.

*Pythium aphanidermatum* is another commonly occurring species which causes "damping off" seedling and root rot disease of sugar beets, radish and other higher plants. Fruit rot of and other plants in also caused by this fungi it forms a white, cottony, fluffy mycelium and cause the rotting. Most of the species of *Pythium* have a very wide host range.

**Vegetative Structure:** Mycelium is irregularly branched, cylindrical, coenocytic and hyaline during young stage. In the older hyphae septa develops irregularly to separate the empty part from the protoplasmic portion, to delimit the mature antheridia, oogonia and zoosporangia from the rest of the hypha. In parasitic forms the mycelium usually is intercellular but may also at times be intercellular with haustoria penetrating the host cells.

**Asexual reproduction:** Asexual reproduction in *Pythium* occurs by means of zoospores that develop within a sporangium. Two principal types of sporangia may be observed. (*i*) spherical or ovoid type; (*ii*) the filamentous type including lobulate type. They may be either terminal or intercalary.

(*i*) **Spherical or ovoid type of sporangium:** If occurs in *P. debaryanum.* Sporangium develops at the tip of a terminal branch and is generally globose. The protoplasmic contents from the hyphae flow into the sporangium until it has grown to its full size, after which a septum is laid down between the sporangium and the basal hyphae. Later on the sporangium apically, forms a beak of variable dimensions. The tip of the beak softens and dissolves to form an exit tube, out of which develops a spherical vesicle into which the multi-nucleate protoplast of the sporangium flows in. Due to protoplast cleavage, zoospores are formed in large number with in the vesicle. All the zoospores move with in the vesicle, initially sluggishly and then with increasing rapidity. After sometime vesicle ruptures and the zoospores are liberated. Zoospores are kidney shaped with two laterally inserted flagella on the concave side. These swim about actively for sometime come to rest, encyst and then germinate by a germ tube.

(*ii*) **Filamentous and lobulated type of sporangia:** Filamentous sporangia is indistinguishable from the vegetative hyphae and it is simple in structure also. It may be terminal or intercalary in position, but it is cut off from the supporting hypha by a septum and are recognisable usually by their denser cytoplasmic contents. Zoospores are formed in the usual manner by cleavage of the protoplast. In *P. aphanidermatum* the lobulated masses develop long emission tubes, at the tip of which is formed a vesicle within which the zoospores are produced and liberated in the usual manner by rupture of its membrane.

During certain environmental condition sporangia behave as conidia and zoospores are not formed from it, germinate by the production of one to six germ tubes. In one species *P. ultimum*, sporangia regularly behave as conidia and in *P. intermedium* sporangia may function like conidia. They are disseminated by wind or water and in the latter they are produced in chains.

**Sexual Reproduction:** *Pythium* is homothallic in nature. Sex organs and sporangia are formed on the same mycelium oogonia and antheridia develop, close to each other and antheridium is present on the lateral side of oogonia. Antheridia and oogonia are monoclinous (develop from the same branch) oogonia is globose and terminal in position sometimes it is lateral in position of the hyphal tip. Intercalary oogonia can also be seen sometimes. In oogonium, nuclei and cytoplasm migrates from the hyphae and delimited by the septa to the supporting hyphae. Mature oogonium contains 10–15 nuclei and undergo mitotic divison. Later on, oospore divides or differentiate into two layer, peripheral layer which is vacuolated is periplasm, and central dense layer is known as ooplasm. In the centre one remains functional nuclei, rest of the nuclei migrates to the periplasm.

Antheridia develops from the same hyphae, or may be from neighbouring hyphae. Antheridia is somewhat elongated or clavate structure. Many antheridia can develop around the oogonia and each contains 4–6 nuclei. Mature antheridia is cut off by the septum from the hyphae. Nuclei under go mitotic divison. One remain functional and rest degenrates. Fertilization tube develops at the point of contact between the two sex organs, by piercing the outer layer of oogonia it enters into the ooplasm. Functional male nucleus by rupturing the tube tip comes out near the egg nucleus. Both the nuclei fuse and the zygote–oospore is formed which develops a thick smooth stratified wall around it.

Two types of oospores may be observed. Oospores that fill the entire oogonial cavity are known as plerotic and those that fill it only partially as aplerotic.

Oospores undergoes a long period of rest where it germinates by germ tube that develops into the mycelium or the germ tube stops growing at a early stage. Protoplast of the oospore migrates through the tip into a vesicle at its top where the zoospores are formed in the usual manner. Kidney shaped biflagellate and uninucleate zoospores are liberated by the rupture of vesicle wall which encyst and germinate into the mycelium first divison of zygote nucleus is meiotic.

***Achlya:*** It is commonly known as watermolds. It is acquatic in nature and occurring in most collections of fresh waters. Mainly saprophytic in habit, it occurs commonly on decaying vegetable and animal matter and also on dead ants and insects. *Achlya* can be easily obtained in laboratory cultures by baiting technique.

**Vegetative structure:** Mycelium is highly branched coenocytic wide and stout. Septa is present only to seprate the reproductive structure from the other hyphae. Cell wall is made up of cellulose.

**Asexual Reproduction:** Asexual reproduction takes place by means of zoospores formed in the zoosporangia. In *Achlya*, the hyphae bearing zoosporangia grow sympodially, giving rise to a number of zoosporangia, but there is no proliferation. Young zoosporangium develops terminally as an elongate structure, is full of dense granular protoplasm with a central axial vacuole and is soon cut off from the remaining portion of the hyphae by a septum. As the development proceeds. Cleavage line develops in the protoplasm extending from the vacuole outwards and dividing the contents into a number of polygonal uninucleate masses. Each of these later on round up, contract and form a zoospore. These are the primary zoospores that are pyriform, with two anterior flagella. Zoospores are not motile out side the zoosporangium and encyst just at its mouth immediately after its emergence and forms a typical cluster of encysted zoospores with in minutes, these encysted zoospores germinate to give rise to the secondary zoospores that are reinform with two laterally inserted flagella. These swim for sometime, encyst and then germinate by a germ tube which develops into a new mycelium. The secondary zoospores of *Achlya* may swim and encyst several times before the cilia are finally withdrawn. This phenomenon is known as repeated emergence. New zoosporangia generally arise as lateral branches beneath the old ones, pushing them aside (Fig. 8.8 A-K).

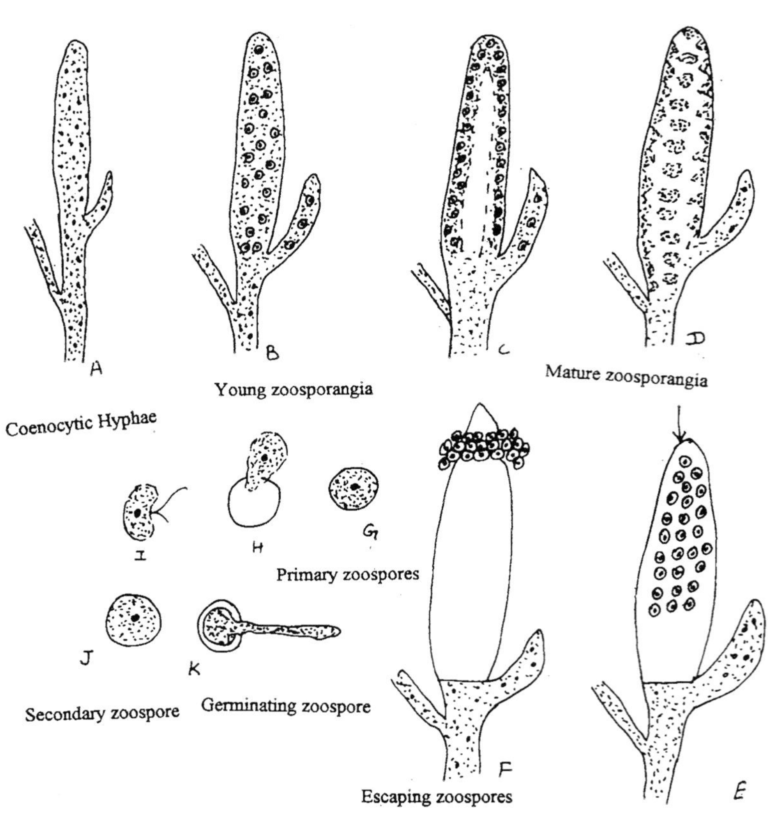


**Fig. 8.8:** Asexual reproduction of *Achlya*

A. coenocytic hyphae. B. Sporangial initial with multinucleate protoplasm. C. Appearance of vacoule and formation of septa. D. Development of polygonal masses. E. Homogenous stage. F. Zoosporangium with uniuncleate daughter protplast. G. Encysted primary zoospores. H. Germination of primary zoospore. I. Secondary zoospore. J. Encysted secondary zoospore. K. Germinating secondary zoospore.

Chlamydospores are occasionally produced in *Achlya*. They are widely in shape and size from spherical to elongated, cylindrical, pyriform or of any other shape. These germinate in the usual way by putting forth a germ tube which may either grow into a mycelium or may develop into a short stalked zoosporangium.

**Sexual Reproduction:** Most species of *Achlya* are homothallic–both the antheridia and oogonia occur on the same mycelium. Some species like *A. ambisexualis* and *A. bisexualis* are heterothallic. In homothallic forms *A. polyandra* and *A. americana* etc. oogonia arise as terminal globose structure filled with dense contents and one soon cut off from the underlying hypha by a septum. The contents of the mature oogonium are differentiated into a single or a number of globular eggs which are initially multinucleate but finally become uninucleate. In the meantime, multinucleate and elongate antheridia also arise on the same hypha that bears the oogonium. One or more antheridia get attached to the oogonium and pierce through its wall by fertilisation tube which may branch with in the oogonium. Each branch reaching an egg where the wall of the tube ruptures and liberates the male of the tube ruptures and liberates the male nucleus. Fusion occurs between the male and the egg nuclei. A thick wall is secreted round the oospores and they undergo a period of rest (Fig. 8.9 A-H).

The heterothallic species *i.e. A. ambisexualis* and *A. bisexualis* require two different types of mycelia–one male another female–for the sexual reproduction Raper (1939) reported that strong male and strong female cultures of *A. bisexualis* and *A. ambisexualis* are placed on the opposite sides of a petridish of suitable agar medium. The strain grow towards one another and the sexual reproduction occurs with the following series of events.

(*i*) Male strain produces antheridial branches.

(*ii*) Female strain produces the primordia of the oogonia.

(*iii*) The antheridial branches grow towards the oogonial initials and wrap around them and antheridia are delimited by septa in the antheridial branches at their tip.

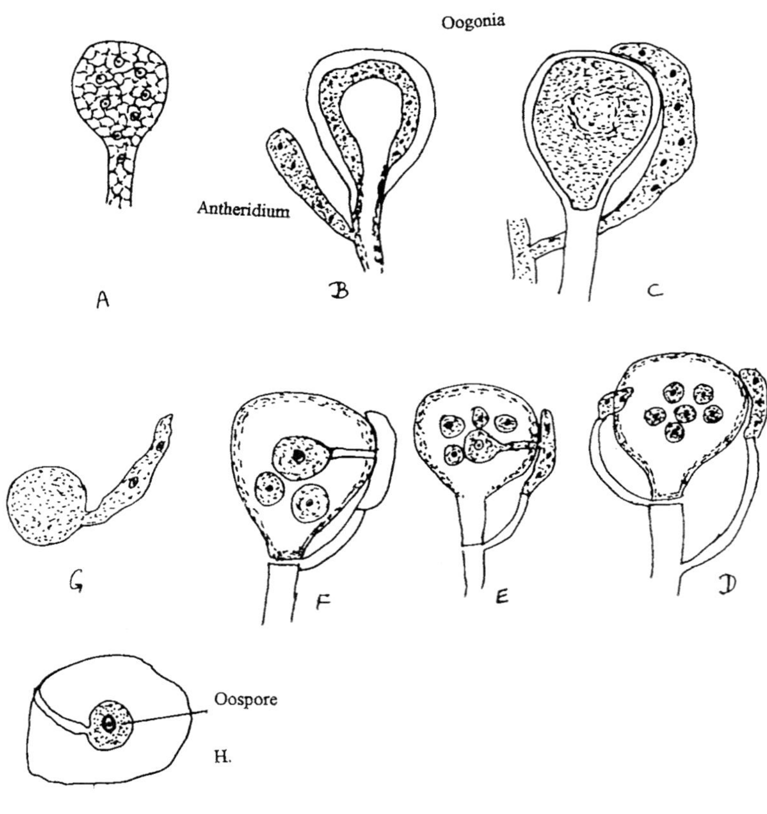


**Fig 8.9:** Sexual reproduction in *Achlya*

A. Multinuclate oogonial initial. B. Anthredium arising from the oogonial stalk. C. Oogonium bearing anthredium. D. Oogoonium bearing two anthredia on the stalk of oogonium. E. Fertilization tube formed by the anthredium. F. Portion of oogonium showing oospheres and remains of anthredium attached to its walls. G. Union of male and female nuclei. H. Mature oospore.

(*iv*) The oogonial initials are cut off from their subtending hyphae by septa to become oogonia.

(*v*) Oospheres are delimited, followed by fertilization of the oospheres and oospore development.

This indicates that the process is controlled by a series of hormones, a hormone is produced by a strain, eliciting a response in the other. Which then produces a different hormone which further elicites its characteristic response in the first strain and so on.

Raper labelled as Hormone A, a substance produced by female strain to induced antheridial branches in the male strain. Hormone B, for one, by which male strain produces a substance which induces female strain to produce oogonial initials. Further direct growth antheridial branches towards oogonial initial by Hormone C produced by oogonial initial and Hormone D produced by delimited antheridia to differentiate oospheres in oogonia.

Hyphal tip growth and chemotropism in both strains is due to phenyl alanine and methionine. Fuosterol is a basic steroid which is present in the fungus and allows to synthesize antheridial and oogonial. It is also confirmed that reproduction in the homothallic *Achlyas* is controlled by the same hormone as it is the heterothallics. The behaviour of the various strains depends on the hormones which they produce and to which they respond.

***Saprolegnia:*** The genus is of world wide occurrence, represented by 20 species. This genus is saprophytic in nature. *Saprolegnia* is present on the hemp seeds, dead ants and insects or other animals are added as baits into a sample of shallow pond water. Baits are surrounded by a long colourless stout hyphae. Hyphae enters into the body of animal tissue. Two species of *Saprolegnia i.e. S. parasitica* and *S. ferax* are parasitic.

**Vegetative Structure:** Mycelium consists of simple or branched, long coenocytic hyphae of wide diameters. Septa are formed only to delimit the reproductive organs stout hyphae project out from the substratum and are clearly visible to the naked eye. Large aseptate hyphae have a multinucleate layer of cytoplasm surrounding a central vacuole, and the cell wall gives good reaction for cellulose.

**Asexual Reproduction**

1. By fragmentation: Fragments regenerate to form the mycelium.

2. By chlamydospores (Gemmae): Asexual reproductive bodies formed in some species of *Saprolegnia*. They are formed by accumulation of hyphal contents in certain portions with subsequent development of septa while separate from the rest of the hyphae. The chlamydospores are of variable shape and size and are generally terminal either occurring singly or in chains. They seprate out from the parent hyphae after maturing and germinate by putting forth a germtube which either grows into a mycelium or develops into a short stalked sporangium forming zoospores in the usual manner.

**By zoospores:** They are produced with in long cylindrical, club shaped zoosporangia that develop terminally on hyphal branches. Zoosporangia develop by the accumulation of dense granular multinuleate protoplasm towards the end of a hyphal branch this portion becomes greater in diameter then the remaining part. Young zoosporangium is cut off from underlying hyphae by means of a septum. There is a central vacuole with in the dense contents of the zoosporanium. The zoospore formation starts by cleavaging of the protoplast which initiates from the vacuole outwards cutting off irregularly polygonal, uninucleate masses of naked protoplasm. These portions later round off develop two anterior flagella and become the zoospores which are pyriform. These are primary zoospores (Fig 8.10 A-L).

Zoospores escape one by one through an apical opening that forms in the softend tip of the mature zoosporangium. The pyriform zoospores move actively for sometime, come to rest, round up and a thin wall is secreted round them, they encyst. After a period of rest, the cysts breaks and the zoospore, instead of germinating by a germ tube, gives rise to another kind of zoospore which is kidney shaped and bears two lateral flagella. These are known as secondary zoospores also swim for sometime encyst and then germinate by putting a germ tube that develops into a new mycelium.

After the escaping of zoospores from the zoosporagium, the under lying hypha again grows through it and develops a new zoosporagium either with in or beyond the original wall. In this way number of zoosporangia sometimes five or six, may develop by proliferation in a single zoosporangium (Fig. 8.10).

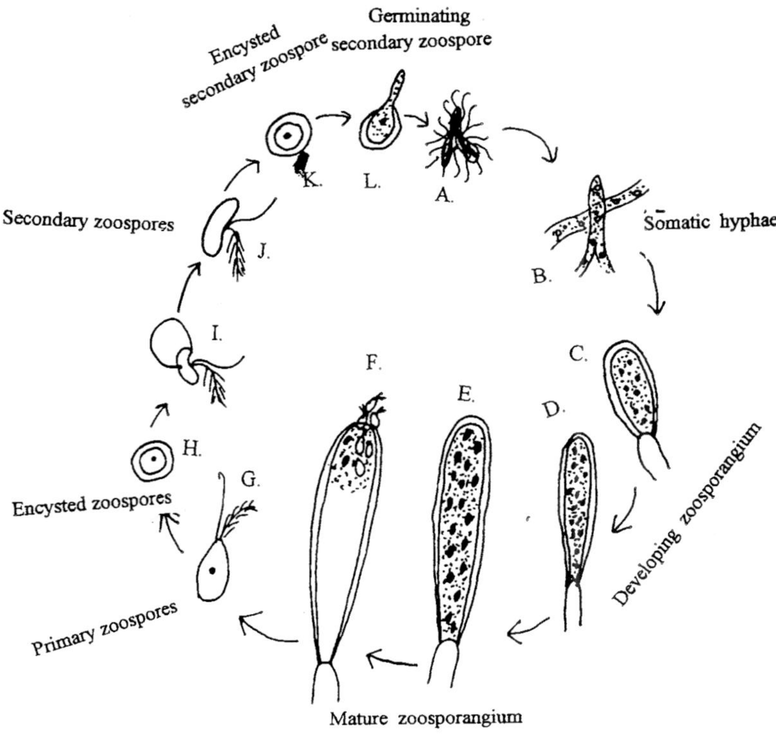


**Fig. 8.10:** Stages in Asexual reproduction

**Sexual Reproduction:** Sex organs are formed on the same mycelium, on which the zoospores are formed.

Oogonia first to be formed, develop on the main or lateral hyphal branches as terminal swellings, usually globose and later separating from the main hyphae by a septum. Sometimes a chain of oogonia may be formed. The multinucleate protoplasm inside the oogonium, is uniformly granular but soon clevaging divides the protoplasm into a number of units. These eggs are multinucleate but later one nuclei remains rest degenrates. Simultaneously, with the development of oospheres either from the same hyphae that bears the oogonium or from neighbouring branches develop elongated, club shaped antheridia. These are multinucleate structure and soon seprated by a septum from the supporting hyphae. On reaching the oogonium these get attached to the oogonial wall and become flattened. One or more antheridia may develop round each oogonium. As the eggs mature, the antheridia push a conjugation tube through the oogonial wall which reaches an egg either directly or may branch out so as to reach several eggs. On reaching the egg, the tube bursts and a male nucleus is discharged into each egg which fuses with the egg nucleus and the osspore develops thick wall. Usually many oospore are formed as there are eggs in the oogonium synzoospores. In many cases there is no fertilization and the oospores may develop parthenogenetically (Fig. 8.11).

Oospores either remain with in the parent oogonium and germinate after a period of rest or they are set free on disintegration of the oogonial wall and lie buried in the mud. After a resting period of several months. They germinate by a germ tube which may develop into the mycelium or may form a zoosporangium. Meiosis occurs at the germination of the oospore.

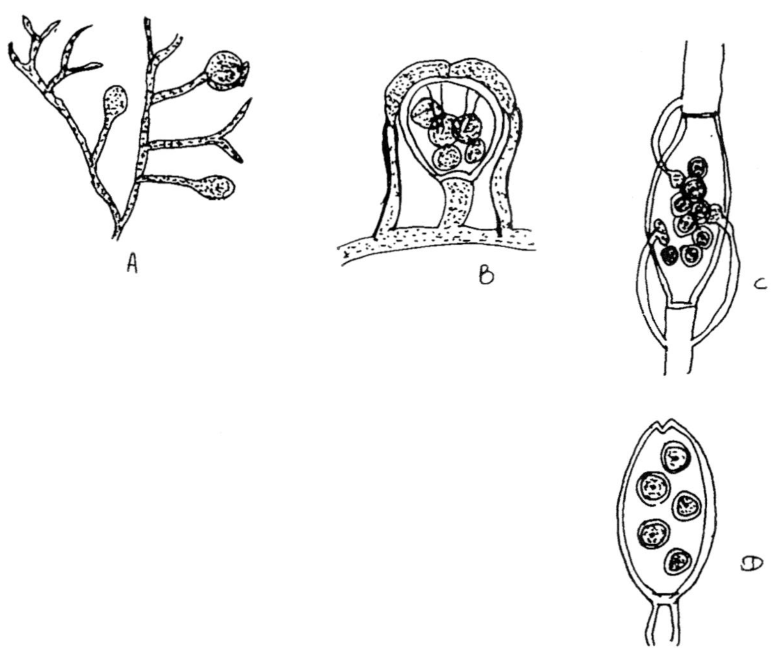


**Fig. 8.11:** Sexual reproduction in *Saprolegnia*

A. A part of the mycelium with young sex organs. B. Terminal oogonium with fertilizaiton tube growing from anthocedia C. Intercalary oogonium with numerous oospheres D. Oogonium with oospore (Synzoospore)

# 9. ZYGOMYCETES

Division Zygomycotina consists of two classes i.e. Zygomycetes and Trichomycetes. The trichomycetes, of which about 150 species are known are obligate parasites that live in the gut of insects and other arthropods. There are about 700 species of Zygomycetes, and members of the order Mucorales are very wide spread and abundant.

The term zygomycetes refers to the production of a thick walled resting spore called a zygospore (zygos — yoke + spore — seed, spore) that develops within a zygosporangium formed as a result of the complete fusion of two equal or unequal gametangia. These gametangia may arise from the same mycelium or from different mycelia. Most zygomycetes produce a well developed mycelium consisting of coenocytic hyphae motile cells are absent in zygomycetes. Asexual reproduction is typically by sporangiospores although some species produce chlamydospores : budding is known to occur under certain conditions in others.

Zygomycetes range all the way from saprobes, through facultative, weak parasites of plants, to specialized parasites of animals and to obligate parasites of fungi including other zygomycetes. Some fungi are used in the fermentation of food items while others are used commercially to produce certain enzymes, acids and so on. Some zygomycetes are also human pathogens, while others are important mycorrhizal fungi. The phylum zygomycota consists of the two class zygomycetes and trichomycetes. Trichomycetes a group of fungi that are ecologically and morphologically distinct from all other fungi.

The principal characteristic that distinguishes class zygomycetes is the production of a thick resting spore called a zygospore (zygos = yoke + spora = seed). The zygospore develops with in a zygosporangium that is formed after fusion of two gametangia. These fungi include coenocytic mycelium, asexual reproduction usually by sporangiospores and absence of flagellate cells and centrioles. The walls of hyphae are composed of chitin, chitosan and polyglucuronic acid. In other forms, however the mycelium is very much reduced and may have more or less regularly spaced septa. Some species have the capacity to grow either as mycelia or yeasts are said to be dimorphic.

Asexual reproduction is zygomycetes may be by means of sporangiospores and some may produce chlamydospores. Oidia and arthospores in addition. Sporangium of zygomycetes is relatively large, usually columellate structure that is borne terminally on a specialized hypha termed a sporangiophore.

Spores of zygomycetes are formed by four fundamentally different processes. The first is known to be sexual in many instances, and the resulting zygospores are formed in a variety of ways following the union of morphologically differentiated and undifferentiated cells or hyphae. The other three processes are asexual. Two of these result in the delimitation of endogenous spores (sporangiospores and chlamydospores) whereas in others, the spores (arthrospores) are formed exogenously.

Zygospores are formed only in zygomycetes due to the sexual reproduction. They are regularly found only in homothallic members of the class but also found in heterothallic species also sexual or zygosporic, state alone cannot be used for establishing botanical taxa in zygomycetes, the predominant asexual, anamorphic–that producing sporangia and sporangiospores–is the one on which the name of species are based and taxa classified in all members of this class.

**Sporangiospore:** It is a spore formed in a sporangium (Ainsworth 1971) and in the Eumycota, it is the characteristic asexual spore of the mastigomycotina and zygomycotina. Sporangiospores are distinguished from other asexual spores formed by members of these subdivisions in being cleaved out of the cytoplasmic contents of the sporangium, without the wall of the sporangium being directly involved in the formation of the spore wall they are enterothallic.

In many zygomycetes the primary asexual spores *i.e.* sporangiospores, develop on specialized simple or branched, usually aerial sporophores, and are formed singly or in rows of few to many units rather than in well defined, more or less globose sporangia or sporangiola as in the majority of mucorales. Sporangial wall being either ephemeral or so closely appressed to the spore wall that it cannot be readily observed by the usual methods.

In 1859, de Bary introduced the term chlamydospores to describe the thick walled spores formed in abundance in hyphal cells on and in the carpophores of *Nyctalis aesteropohora.*

Van Tieghem & Le Monnier (1873) adopted de Bary's term to designate one of the two forms of asexual spores they recognized in the mucorales–forme sporangale, and forme chlamyde. They defined the chlamydospore as a spore.

(1) formed by the same mycelium giving rise to sporangia but developing endogenously by local condensation and walling off of protoplasm.

(2) Liberated by the break down of the enveloping hyphal membrane. Ainsworth (1971) defined chlamydospores as a thick walled, non deciduous, intercalary or terminal asexual spore made by the rounding up of a cell or cells 'gemma'. Formation of endogenous chlamydospores is common in many zygomycetes, formation of truly exogenous spores *i.e.* conidia, is rarely encountered and may be represented only by the production of chains of variable elongate, ovoid or globose spores commonly formed by the submerged hyphae of several species of *Mucor micheli* growing in culture. Such spores have called oidia or gemmae. They appear to be formed in basipetal succession, and they may proliferate by yeast like budding.

Gemmae in the zygomycetes are best represented by the yeast or bud cells formed in culture by arthrospores and especially sporangiospores of species of several genera of Mucorales (*Actinomucor, Mucor, Mycotypha*).

Zygomycetes are divided into three order:- Mucorales,. Entomophthorales, Zoopagales.

**Mucorales:** They are widely distributed fungi, great majority of the species are saprophytic, occur on organic subtrata, whilest the remainder are parasitic on other members of the order. Typical colonies consist of coarse hyphae growing loosely while in the early stages of growth and becoming grey or brownish with the production of fruiting structures. In the older mycelium, especially in the aerial portions, septa may divide it into pluri nucleate segments, but the young mycelium and that submerged in the substratum usually remain nonseptate cell wall contain chitin and pectose and no cellulose.

Asexual reproduction is by the formation of non motile, encysted spores in sporangia terminal to the hyphae. These sporangia are formed in the same manner as in Perenosporales. Passage of a portion of the contents of the hypha into a terminal enlargement which is then cut off from the hypha by a septum. Multinuclear content of the sporangium are divided by cleavage planes into many naked at first, polyhederal cells containing one or more nuclei each. These then round up and encyst and escape by the rupture of the dissolution of the sporangium wall. They germinate by the germ tube. In most genera the tip of the sporangiophore is swollen, the swollen end columella projecting into the sporangium. The columella may be of various shapes–globose, ovoid, hemispherical etc. Sporangial wall may be thin, when the spores are liberated by its rupture or dissolution, or may be cutinized and shot off, or broken off in one piece. Various modifications occurs of the typical globose, many spored sporangium.

**Habitat:** They are mainly saprophytic on vegetable matter, more rarely on animal matter, and are abundant in the soil and in plant debris. Many are coprophilous. Some are weak parasites on living plant tissues which are rich in stored food but not active such as the roots of the sweet potato. A number of species are parasitic upon other fungi, even upon other mucorales. *Piptocephalis*.

The order Mucorales is classified into 11 families. Thaminidiaceae. Cunnighamellaceae, Choanephoraceae, Pilobolaceae, Mortierellaceae, Endogonaceae, Syncephalastraceae, Piptocephalidaceae, Dimargaritaceae and Kickxellaceae.

**Family Mucoraceae:** Mucoraceae is one of the largest families of Mucorales. It contains *Actinomucor, Mucor, Rhizomucor, Parasitella, Zygorynchus* and *Circinomucor*. All of these genera produce non apophysate sporangia with persistent walls. Slight constriction of the sporangiophore immediately below the sporangium. Some taxa produces stolons and rhizoids. Zygospores have opposed, non-appendaged suspensors.

*Mucor ramosissimus* and species of *Rhizomucor* are known to cause disease in humans, Mucorales are industrially important organisms. They are isolated from dung, plant material and the air.

***Zygorynchus:*** The genus is closely related to the *Mucor*. It forms many spored sporangia of similar structure of *Mucor*, but differ that all the species are homothallic, suspensors of the zygospores are unequal in size. Two species are isolated fairly frequently in this country *i.e. Z. moelleri* and *Z. heterogamous*. Spores are round, zygospores black, very long, varying in size from 45 m to 150 m, chlamydospores formed in the mycelium.

**Family Absidiaceae**

***Absidia:*** This genus differ in several respects from *Rhizopus*. The rhizoids and stolons are not so clearly differentiated, the sporangiophores arise from the stolons and not from the points of attachment of the rhizoids, sporangia are relatively small and pear shaped, most characteristic feature in a well marked 'apophysis' funnel shaped base to the sporangium, where the walls of sporangium and columella are united. Zygospores when present are surrounded by coarse hairy out growths from one or both suspensors, this feature arising to place at once the homothallic species of the genus. They are thermophillic with optimum temperature for growth close to 37°C. It is pathogenic to human beings. *A. corymbifera, A. spinosa* and *A. ramosa* are the common species of the *Absidia* (Fig. 9.1 J-K).

**Family Phycomycetaceae**

***Phycomyces:*** This genus is readily recognized by their characteristic sporangiophores very long, stiff, and with metallic sheen, looking like oxidized steel wire. Four species are known but only two are of any importance. Sporangiophore is upto 20 cm in height. Sporangia black, 500 m in the middle, spores elongated ovate. Common species are *P. nitens* and *P. blakesleeanus*.

**Family Thamnidiaceae**

***Thamnidium:*** Main sporangiophores bear lateral clusters of sporangioles as well as large terminal sporangia, the sporangioles resemble miniature sporangia, containing from two to a dozen or so spores and are formed richly branched out growths from near the base of the sporangiophore. Two types of sporangium are present. The large terminal sporangium with a columella, and the small lateral separable sporangioles which lack a columella. The zygospores are formed on approximately equal suspensors as in *Mucor*. In *Helicostylum* the sporangioles are borne on short circinate branchlets from the unforked lateral branches. In *Dicranophora* the sporangioles are one to few spored and have a rounded columella and thin sporangial wall. Spores are very variable in size and mostly bean shaped, they are mostly ellipsoid and more numerous and on the whole smaller in the terminal sporangium.

**Family Choanephoraceae**

Large sporangia are produced in two genera and possess a columella. The sporangioles instead of arising singly at the tips of forked branches are found crowded on the surface of the swollen apical portion of a large sporangiophore of its branches. They are monosporous and indehiscient in three genera and several spored and dehiscent in one genus–*Blackeslee*. Most of the species are parasitic or saprophytic on flowers or other vegetable matter. *Choanephora* produces sporangia and indehiscent monosporous sporangioles, while *Cunnighamella* produces only latter (Fig. 9.1 N-O).

**Family Syncephalastraceae**

***Syncephalastrum:*** Spores are formed in long tubular sporangia radiating from a swelling on the end of sporangiophore. Heads gives the resemblance to the *Aspergillus* at low magnification. At a high magnification the chains of spores are seen to be enclosed in tubular membranes. Syncephalastraceae contains a single genus

*Syncephalastrum.* Characterized by the formation of branched sporangiophores that bear apical fertile vesicles covered with merosporangia. Zygospores have pigmented ornamented zygosporangium walls and opposed suspensors. Merosporangia are multispored in *S. racemosum*, while unispored in *S. monosporum*. They can be isolated from dung and soil.

**Family Pilobolaceae:** Sporangia more or less flattened vertically with a thick dark coloured apical wall which does not break or dissolve sporangiophore is swollen at the base *Pilobolus* or a slight one subtending the sporangium. Columella is more or less conical or may project almost to the top of the sporangium. In *Pilobolus* when the maximum osmotic pressure has been attained in the subsporangial vesicle it ruptures in a circular slit so that the columella and adhering sporangium are violentily expelled by the mass of liquid forced out by the contraction of the ruptured vesicle. The dung inhabiting species of *Pilobolus* are very easily obtained by bringing freshly dropped horse manure into the lab and enclosing it in a dish. In *Pilobolus* the mature sporangiophore bends towards light as a result of the subsporangial vesicle acting as a lens to focus transmitted light, to a spot beneath the vesicle. The subsporangial vesicle has a high turgor pressure, and when it eventually ruptures, it ejects the sporangium toward the source of light. The sporangium adheres to the surface it contacts. On a wet surface the calcium oxalate crystals on the sporangium wall create an unwetted surface that causes it to rotate (Fig. 9.1).

*Pilaria* and *Utharomyces* are less well known members of this family.

**Family Mycotyphaceae:** This family includes the genus *Mycotypha.* It is characterized by the formation of circumscissile zone of dehiscence at the function of the pedicel and denticle, which facilitates sporangiolum and pedicel release, and by the ready formation of yeast cells on the surface of nutrient rich culture media. *Mycotypha* forms cylindrical, fertile vesicles with dimorphic, unispored sporangiola, attached by denticles zygospores have opposed suspensors.

**Family Gilbertellaceae:** This contains single genus *Gilber tella persicaria*, which causes a storage rot of peaches, nectaries, tomatoes and pears. Sporangiphore arise directly from the substrate and terminate in dark columellate sporangia. Spores are usually broadly fusiform, smooth walled and hyaline with several long, thin appendages arising from the apices. Calcium Oxalate crystals are especially abundant in *G. persicaria*. Zygospores have an ornamented zygosporangial wall and opposed suspensors.

**Family Piptocephalidaceae:** The species of this family are largely parasitic on other mucorales although some are saprophytic. They sporangia are narrow and more or less clavate or cylindrical with the spores usually in one row, often appearing when mature like chains of conidia. The numbers of spores formed in the sporangium varies from 2 to 30. Sporangium breaks into monosporous segments, the spore being enclosed in a sporangial wall.

**Family Dicranophoraceae:** Dicranophoraceae contains four genera. *Sporodiniella* parasitize tropical insects, other three genera *Dicranophora, Spinellus, Syzygites* are necrotrophic parasities of mushrooms. These organisms produce large apophysate sporangia. *Dicranophora* and *Spinellus* grows at low temp, while *Syzygites* and *Sporodiniella* grow well at higher temperatures. Zygospores have opposed suspensors.

**Family Saksenaeaceae:** It have single genus *Saksenea vasiformis*. This genus produces at flask shaped columellate sporangium with a long neck. Stolons and rhizoids are present, and the sporangia are formed singly or in pairs from a dichotomously branched sporangiophore, zygospores are not known.

**Family Sigmoideomycetaceae:** Sigmoideomycetaceae contains two genera, *Sigmoideomyces* and *Thamnocephalis*. These are mycoparasites, they donot form haustoria but produce inflated contact organs, through which host nutrients are absorbed. They are regularly septate, dichotomously branched fertile hyphae that produces pairs of stalked fertile vesicles from the cells at the branching points as well as sterile spores. They are found on dung, rotting wood, leaves and other organic material. Zygospores are unknown in the family.

### *Mucor*

**Habit and Occurrence:** *Mucor* offers the typical example of the family Mucoraceae. About 60 species of the genus are known so far.

It is world wide in distribution occurring commonly as a saprohytic mould particularly during the rainy season or in humid conditions on varied substrate like, bread, jellies, syrups, food stuffs, rich in sugar, leather, dung and other decaying vegetable matter. It is commonly called as the bread mould because of its easy occurrence on stale and moist bread or as the pin mold because of its long pin head like sporangiophores bearing terminally globular sporangia. Some species of *M. mucedo*, *M. racemosus* are common air contaminants. While others like *M. flavus, M. hiemalis, M. strictus* are found in soil. *Mucor javanicus* is an industrially important species being used for alcoholic fermentation and *M. pusillus* has medical significance being the cause of a disease in man known as the mycosis of the internal organs.

The thallus forms a white cottony, mycelium on the substratum. Mould commonly appears on dung in *M. mucedo* and the one on bread is *M. stolonifer*.

**Vegetative structure:** Mycelium consists of freely branched, long and slender hyphae occurring in the form of a cottony growth over the substratum. The hyphae are aseptate, septa occasionally developing in the older aerial hyphae, or during formation of the reproductive stage. The walls are composed of chitin and whether cellulose is also present has been a matter of controversy. The granular protoplasm with in the cell wall contains large number of vacuoles of varying size, oil droplets, glycogen bodies and numerous small nuclei. Protoplasmic streaming can be very well observed in younger and vigorously growing hyphae.

The mycelium growing on substratum could be distinguished into two parts. The hyphae penetrating it serves as absorptive hyphae and the remaining bulk of the aerial hyphae give rise to sporangiophores (Fig. 9.1 F-G).

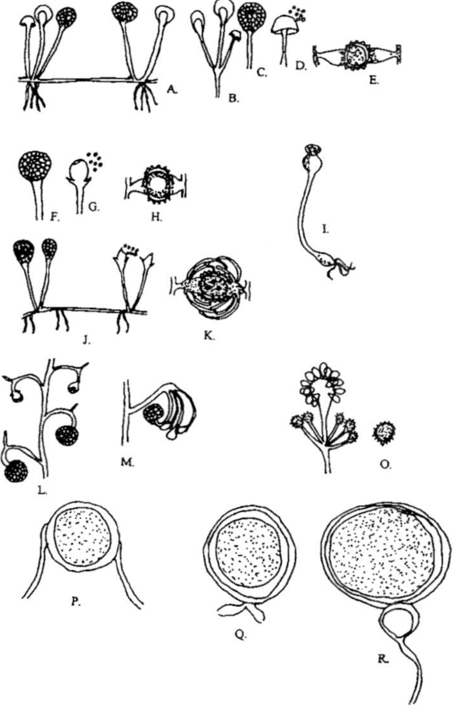


**Fig. 9.1:**

A. *Rhizopus* stolon bearing sporangiophores and rhizoids. B. Sporangiophore with node and branches. C. Sporangiophore with intact sporangium. D. Dehisced sporangium with peltate columella and sporangiospores. E. Zygospore. F. Mucor sporangiophore with intact sporangium. G. Dehisced sporangium showing columella with distnict collar. H. Zygospores. I. *Pilobolus* sporangiophore with trophocyst J. *Absidia* stolons with sporangiophore and rhizoids. K. Zygospore with appendage. L. *Circinella* sporangiophore branch showing sterile spine with collar and columella. M. Sporangiophore showing characteristic sporangial wall. N. *Cunnighamella* sporangiohore bearing sporangia. O. Sporangia. P. Zygospore of VAM fungi attached to undifferentiated hyphae like suspensors. Q.. *Endogone* species. zygospore. R. *Gigaspora calospora*. azygospore.

Some species of *Mucor*, growing on concentrated sugary solutions or in medium rich in nutrients and having high osmotic concentrations, from yeast like growth *i.e.* a chain of bead like cells instead of the normal filamentous mycelium. This is known as torula stage or *Mucor*.

**Asexual Reproduction:** It takes place by four methods:–

1. **Fragmentation:** Each fragment of vegetative hypha regenerate to develop the mycelium.
2. **By Oidia:** They are thin walled structures occurring usually in chains and giving a beaded appearance to the mycelium. In some species of *M. racemosus* under certain conditions particularly when the fungus is growing in nutrient rich media as sugar solutions, the hyphae break up in small portions which round up and form oidia. Each of these on germination, is capable of giving rise to new mycelium.
3. **By Chlamydospores:** They are thick walled, structure meant to serve as resting bodies. They arise endogenously with in the parent hyphal wall. The contents of the hyphae break up into multinucleate protoplasmic portions of varying size which round up and develop a thick wall around them, due to which they can with stand unfavourable conditions. Under suitable conditions these again form the new mycelium.
4. **By Spores:** Spores are present in the sporangium this is the common and characteristic method of asexual reproduction in *Mucor*. The sporangia occur as apical or terminal swellings on aerial, vertical sporangiophores which may be simple as in most species of *Mucor* or may be branched each branch terminating in a sporangium as in *M. racemosus* or *M. brunneus*.

The apical tip of the aerial hyphae branches, destined to become sporangiophores, swell up due to the continuous flow of cytoplasm, nuclei and reserve food material from the underlying hyphae. They gradually enlarge with further flow of contents resulting in the formation of a large globose structure. This is the young sporangium as it matures, its contents get differentiated into a thick dense layer of cytoplasm with large number of nuclei towards the centre. On the border of these two regions develops a foamy protoplasmic layer in which a number of narrow, flattened vacoules make their appearance. These vacuoles gradually enlarge and coalesce laterally with each other forming a continuous cleavage cavity between the denser peripheral and the vacuolated central portions of the sporangial contents. A wall is recreated toward the inner side of the cavity. The central vacuolated portion bound by this wall or septum is called the columella. The nuclei of this portion gradually degenerate.

In the contents of the outer fertile portion, cleavaging takes place dividing the peripheral mass into many irregular units of protoplasm each unit containing 2 to 10 nuclei and cytoplasm. These units round off and recreate a wall around them, thus forming the spores known as aplanospores. Gradually, as the spore mature, the thin wall of the sporangium dries up and becomes fragile. With the slightest disturbance of air currents, it gets ruptured and the spores are liberated. The sporangiophore and the nearly globular columella persist even after the liberation of the spores sometimes remnants of the sporangial wall may be observed forming a basal collar at the base of the columella.

The spores are dark coloured and usually elliptic to ovoid in shape. Their dark colour is on account of the relatively thick wall that is recreated round them and this makes them withstand unfavourable conditions also. Falling on suitable substratum, in presence of proper moisture and temperature, the spores germinate by putting the germ tube that develops into a new mycelium.

**Sexual Reproduction:** Sexual reproduction in *Mucor* takes place by formation of two multinucleate gametangia which on fusion gives rise to zygospore. The zygospore formation in various species of *Mucor* was known for a long time. Sexual behaviour is grouped in two categories. (*i*) homothallic species (*ii*) heterothallic species (Fig. 9.2).

**Homothallic species:** Which could produce zygospores on a single thallus *i.e.* on mycelium developing from a single spore (Fig. 9.3).

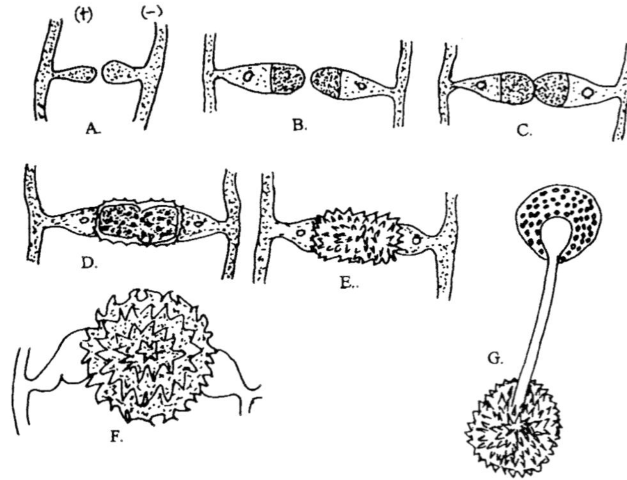


**Fig. 9.2:** Sexual reproduction in *Mucor*. A-G stages in the sexual reproduction. A two hyphae of opposite strains giving rise to progametangia. Fusion of two gametangia resulting in the formation of thick and warted zygospore.

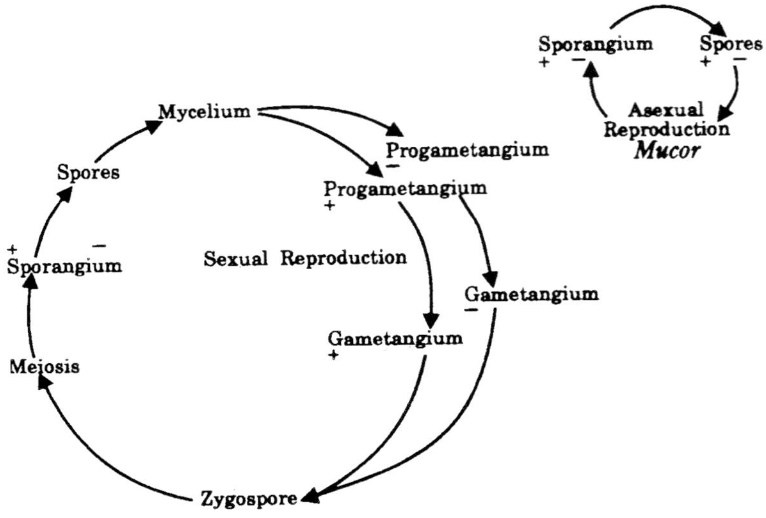


**Fig. 9.3:** Graphic life cycle of homothallic species *Mucor genevensis*.

**Heterothallic species:** Which required two compatible thalli to form zygospores. Since these thalli differ physiologically could not be distinguished morphologically, they were designated as (+) or (–) strains. Former is female and later is male (Fig. 9.4).

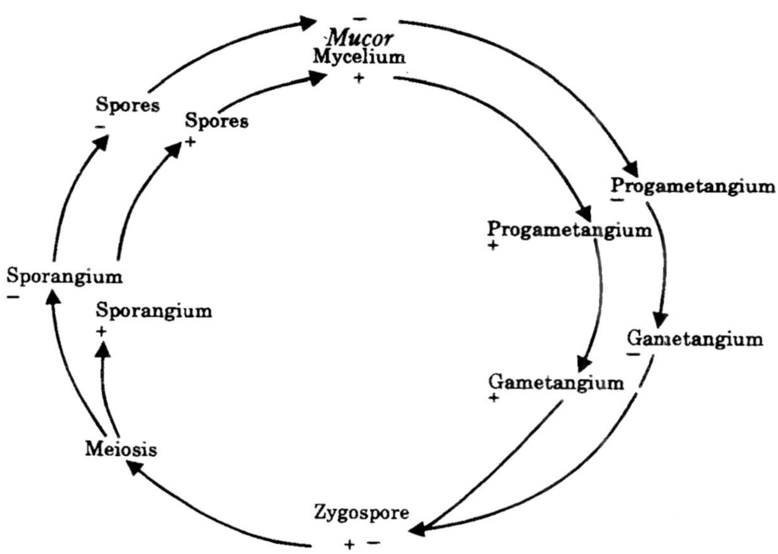


**Fig. 9.4:** Sexual Reproduction in a heterothallic species *Mucor mucedo*

Most of the species of *Mucor* like *M. mucedo* and *M. hiemalis* are heterothallic these are some homothallic species also *e.g. M. tenuis, M. genevensis*. However the formation and development of gametangia and the stages that follow are similar in both of these except, that a single thallus would not develop zygospores in heterothallic species.

In the heterothallic species, the sexual reproduction is initiated when hyphal branches (+) and (–) strains comes in contact with one another. Swelling occurs at the tip of the copulating branches at the point of contact and this develops to form a club shaped structure the progametangium. Dense cytoplasm and many nuclei flow towards the contracting tips which enlarge further. A septum is laid down separating the terminal portion which is termed as gametangium. The remaining part of the programetangium is called as suspensor. The undifferentiated multinucleate protoplast of each gametangium is also termed as a gamete or coenogamete. Both the gametangia and suspensors are approximately equal. As the gametangia mature the separating wall dissolves from the middle outwards and intermingling of the contents of the two gametangia takes place followed by nuclear pairing and fusion one (+) and another (–) giving rise to large number of diploid nuclei. The nuclei which do not get a mate, probably degenerate. The young zygospore lying with in the parent gametangial wall, enlarges considerably and secretes several layers of thick wall around it. As the zygospore mature, it breaks up the original gametangial wall into small pieces that fall apart exposing the outer thick, spiny and dark exospore. The wall of mature zygospore is a five layered structure, two layers in the exospore and three in the endospore. The second layer of the exospore is carbonaceous brown, black and fragile. The zygospore germinates after a long period of rest lasting, one to several months. On germination, the outerwall cracks and out of it emerges a germsporangiophore that terminally bears a germ sporangium, which contains a number of spores. Meiosis probably occurs early in the zygospore (Fig. 9.1 H) so that nuclei in the resting period are haploid or sometimes it may occur during germination of the zygospore. In heterothallic species like *M. mucedo*, and *M. heimalis* all the spores in the germ sporangium belong to only one sexual strain, either (+) or (–). Diploid zygospore under go meiotic division before it undergoes period of rest. Only a few nuclei probably belonging to one sexual strain, remain while the rest degenerate. The remaining nuclei on germination, divide and enter the germsporangium to produce only one type of spores. In the homothallic species all the spores that are produced in the germ sporangium are potentially bisexual. (± both).

Occasionally, failure of gametangial copulation results in parthenogamous development of zygospores which are called azygospores.

***Rhizopus:*** It is another commonly occurring member of the family Mucoraceae. It is represented by 35 species which grow as saprophytes over a wide variety of organic substrate. *Rhizopus* like *Mucor* is also the common bread mold appearing very easily on a most piece of bread if it is left for 2-3 days. *Rhizopus nigricans* causes severe decay of sweet potatoes in storage, while *R. oryzae* is responsible for alcoholic fermentation and some other species cause lactic fermentation.

The vegetative hyphae spread out uniformly with in and upon the substratum. *Rhizopus* resembles *Mucor* is nearly all the essential details concerning the development and stages that follow in the methods of asexual (Fig. 9.5 A-K) and sexual reproduction but it is different in some respects that are described (Fig. 9.5 A, B, C, D, E).

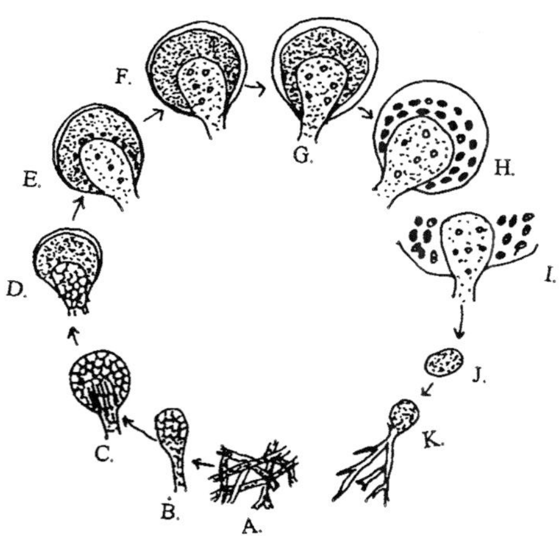


**Fig. 9.5:** Asexual Reproduction in *Rhizopus stolonifer.*

A. Coenocytic hyphae. B. Hyphal tip enlarging to form a sporangium. C. Apical swelling on young sporangiophore. D. Sporangium showing outer and inner part differentiated. E. Small vacoules appearing between inner and outer parts of sporangium. F. Later stage showing enlargement and fusion of vacoules to forms Columellar cleft. G. Late stage showing enlargement and fusion of vacoules to form Columella. H. Mature sporangium. I. Sporangium after. J. Spore. K. Germination of spore to form a new mycelium.

It differs *Mucor* essentially in possessing well differentiated long, stout, creeping runner like hyphae which spread horizontally on the substratum. These hyphae are usual, coenocytic but consists of a distinct node from the under surface of which develop appresoria and rhizoids that branch out penetrating the substratum. From these nodes, arise vertical sporangiophores forming distinct groups. Such hyphae are termed stoloniferous forming stolons. The rest of the details in sporangium development is similar to *Mucor*. *Rhizopus nigricans* is a heterothallic species and *R. sexualis* is homothallic.

Key to common genera of Mucorales

1. Sporangia tubular, radiating from a vesicular swelling –*Syncephalastrum*
2. Many spored sporangia and few spored sporangioles both present –*Thamnidium*
3. Sporangiophores stiff, dark coloured, metallic in appearance –*Phycomyces*
4. Rhizoids and stolons present.

   Rhizoids and stolons absent.
5. Sporangia large, globose, sporangiophores arising from points of attachment of rhizoids –*Rhizopus*
6. Sporangia small, pear shaped, sporangiophores mainly as branches from the stolons –*Absidia*
7. Homothallic zygospores with very unequal suspensors –*Zygorynchus*
8. Homo or Heterothallic, zygospores, when present, with approximately equal suspensors –*Mucor*

**Endogonales and Glomales:** Vam fungi are kept in two orders. Endogonales and Glomales. Endogonales has one family with two genera. VAM/AM fungi are formed by members of Glomales. Glomales consist of two sub order Glomineae and Gigasporineae. The former has two families with four genera and the later has one family with two genera (Mukerji, 1996)

Taxonomy of any group of plants is an important aspect for a botanist (Bursdall, 1990). During the last two decades systematics of Vam fungi has gained significance because of their role in soil fertility, nutrient uptake and biocontrol of plant diseases. Many of the fungi have not yet been cultured axenically which also includes Vam fungi therefore their identification depends on specimens directly isolated from soil on the maximum observable characters.

Benjamin 1979 kept eight genera of endogonaceous fungi under Endogonales in a singly family Endogonaceae. Morton and Benny (1990) divided this order in two indpendent orders Endogonales and Glomales on the basis of their spore structure and mycorrhiza development. The present Endogonales has one family (Fig. 9.1 Q), Endogonaceae with two genera, *Endogone* and *Sclerogone*. These fungi generally develop only arbuscules in plants from temperate parts of the world, but most of the studies from tropical plants definitely report occurrence of vesicles in the root.

Taxonomy of Glomales is based on the structure of their spores/ sporocarps (Morton, 1990) some authors lay more emphasis on the wall structure which also is a valid criterion (Walker, 1983) Glomales and Gigasporineae. The six genera included in this order are placed under three families (Morton and Benny, 1990). In the members of the family Glomaceae *i.e. Glomus* and *Sclerocystis* the spores are formed in sporocarps spores arising in an orderly manner from sterile central plexus. In *Acaulospora* and *Entrophospora* belonging to family Acaulosporaceae, a saccule is formed terminally on a sporogenous hyphae, after which the spores are formed laterally or in between (Fig. 9.1. R).

The spores in *Gigaspora* and *Scutellospora* of family Gigasporaceae are expanded from and borne on a bulbous sporogenous cell and have much larger spores than in other four genera.

Members of Glomales are often referred to as VAM fungi because they form **vesicular-arbuscular mycorrhizae** called as Vam fungi and sometimes known as **endomycorrhizae**. It has been estimated that vesicular arbuscular mycorrhizae can be found in 70% of all plant families. Members of Glomales form mycorrhizal relationships with most agronomically important angiosperms, some gymnosperms as well as certain bryophytes and pteridophytes and even a few algae. Vam fungi do not change the external morphology of the root neither they form any Hartig network. The hyphae grow both between and into the cortical cells by penetrating the wall and causing invagination of the plasma memberane. They produce coils, highly branched haustorium like structure are known as vesicles they can be round or oval in shape. Vesicles are formed either between or within host cell walls and are thoughts to function as energy stores for use by the fungus when the supply of host metabolites is low. Arbuscules are highly branched hyphae that extend through the cell wall, greatly invaginating the host cell plasma membrane. The branches of these specialized hyphae create a large surface area between the fungus and the host cell plasma membrane and appear to be involved in the bidirectional structure of metabolites and nutrients by the two mycorrhizal partners. Arbuscules life span is of very short duration due to their disintegration and digested by the cells of the plant (Fig. 9.6).

VAM fungi provide benefits to their host plants. Vam fungi hyphae extend into the soil away from roots and greatly increase the potential for absorption of water and the uptake of phosphorus and other nutrients by the plant. Vam fungi may absorb and transfer metabolites from other fungi, bacteria, actinomycetes, algae and cynobacteria in the rhizosphere of the associated plants.

The replacement of the pesticides by the controlled use of microorganisms can be considered as a key component in the development of a sustainable agriculture to protect plant against pathogens in an economically profitable manner with a minimized environmental pollution. Mycorrhizal fungi have been reported to the inhibitory effect on plant pathogens and in reducing the disease severity in majority of the studies as neutral or occassional increase in disease severity. Reduction of disease symptoms has been described for fungal pathogens such as *Phytophthora, Fusarium, Pythium, Rhizoctonia, Sclerotium, Verticillium, Aphanomyces, Macrophomina, Phoma, Olpidium and Cylindrocarpon.*

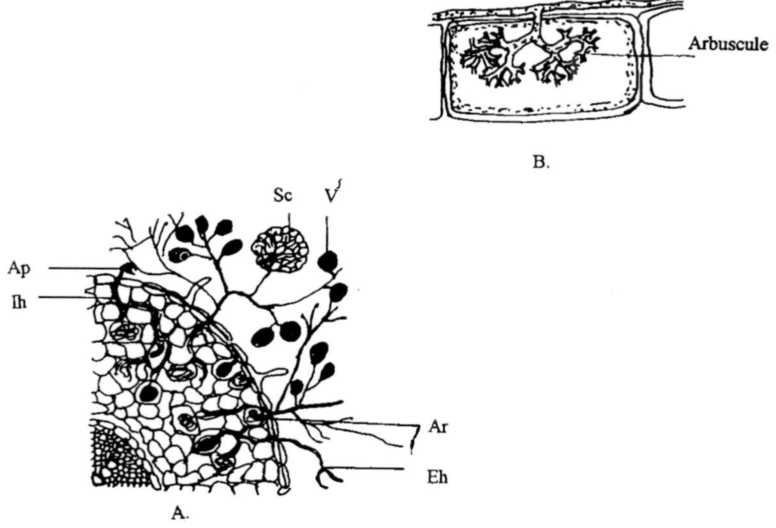


**Fig. 9.6:** Vesicular arbuscular mycorrhizal fungi entering into the root cells

A = Association of Vam fungi in root. B = Detail of absuscular structure inside the cell. Eh = External hyphae. Ap = Appresopium. Ih = Internal hyphae. V = Vesicle Ar = Arbuscule Sc = Sclerocystics

Glomalean species are common soil fungi (Gerdemann and Trappe, 1975), but since most species produce their spores underground. Special techniques usually are required to detect their presence. Although some of the larger sporocarps can be found by randomly raking soil and leaf litter, wet sieving and decanting of soil samples usually are necessary to collect small sporocarps and free spores (Gerdemann and Nicolson, 1963). As these fungi can not be cultured, pot cultures are the only practical methods of maintain these organisms for study (Schenek, 1982, Mukerji et. al. 2002).

**Trichomycetes**

Fungi belonging to the class Trichomycetes are distinct from other fungi i.e. ecologically and morphologically. Members of this class are obligately associated with insects, millepedes, living arthropods and crustaceans. Most species of Trichomycetes grow internally with in the gut of their hosts, only one species occurs on the outer surfaces of arthropods. In many instances, hosts are of acquatic forms. Most of the species found primarily in the hind gut where they attach to the chitinious gut tining by means of a specialized holdfast, while other species gives a fuzzy appearance hence the name Trichomycetes or 'hair fungi' given to the organisms. Fungus absorb their nutrients from the contents of the gut lumen in which they are present. Most common genera of Trichomycetes i.e. *Amoebidium, Asellaria, Harpella, Enterobryus, Genistella, Genistellospora* and *Smittium* etc. Most of the species produce microscopic, limited thalli that depending upon the species may be branched or unbranched. The mycelia of some have regularly occuring septations while reproduction cells.

Asexual reproduction in Trichomycetes may involve amoeboid cells, arthrospores or sporangiospores. In order (Harpellales) special structures known as trichospores are produced. Trichospores are exogenous dehiscent, usually elongate sporangium containing a single uninucleate sporangiospore and having one to several basally attached filamentous appendages continuous with the sporangial wall. There is no direct proof for sexual reproduction in any Trichomycete. Harpellales produce biconical structure referred to as zygospores. These thick walled spores are thought to result from sexual reproduction as they typically form after conjugation between different thalli. Each spore actually is produced on a hyphal branch called the zygosporophore, which arises either from one of the conjugation tube forming between the fusing cells. Moss *et.* al., (1975) distinguished zygospores in Harpellales, all of which are basically biconical in shape.

Because of the presence of zygospores in Harpellales Benjamin (1979) actually included the order in zygomycetes. The possible phylogenetic relationships between these groups are discussed at some length by Moss and Young (1978). Some molecular data hint that the orders of Trichomycetes may not be closely related (Lichtwardt, 1986).

# 10. ASCOMYCETES

Members of the sub division Ascomycotina, commonly referred as the Ascomycetes, are those fungi in which the sexual process involves the production of haploid ascospores through the meiosis of a diploid nucleus in an ascus. Most Ascomycetes also carry out asexual sporulation conidiospores (conidia) being produced on specialized aerial hyphae, the conidiophores rise above the substratum. The sexual phase of an ascomycete is now termed the **telomorph**, and the asexual phase the anamorph.

Many ascomycetes produce their asci in complex fruiting bodies termed ascocarps. Such ascomycetes were formerly regarded as Euascomycetes and classified on the basis of ascocarp form. The Discomycetes were those which had a disc shaped ascocarp the apothecium on which asci are exposed. The pyrenomycetes were those that produced asci with in a flask shaped ascocarp, the perithecium. The Plectomycetes were those in which the asci developed inside an approximately spherical ascocarp, the cleistothecium. Ascocarp, form is crucial in relation to spore dispersal. An apothecium is ideal for the discharge of ascorpores into the air. A perithecium gives some protection but limits the rate at which ascospores can be discharged. In a cleistothecium the asci are well protected, but can be released by the rupture of cleistothecium. There are a very large number of saprotrophic and parasitic ascomycetes, perhaps 15,000 which have ascocarps. In addition there are many more such ascomycetes about 14,000, that have a mutualistic association with phototrophic microorganisms and constitute the lichens. Other ascomycetes formerly termed the Hemiascomycetes (Half ascomycetes) do not have ascocarps, solitary asci being produced. Such ascomycetes are not numerous, but include many important yeasts.

Members of this group lack ascogenous hyphae and ascocarps, and the asci are formed within thick walled ascogenous cells, that have been called cysts or chlamydospores. In *Protomyces* ascus has been interpreted as a 'compound ascus'. Ascospore release in the archiascomycetes may be by forcibly discharge from the ascus in some species of *Taphrina* and *Protomyces* but may be variable even with in these species depending on environmental conditions.

Ascospores are released passively upon rupture of the ascus wall in *Schizosaccharromyces* and *Pneumocystis*. Sexual reproduction is unknown in *Saitoella* chitin is found in the walls of *Taphrina, Protomyces, Schizosaccharromyces* and true yeast. Ploidy level of nuclei in somatic structure varies in various genera. In *Taphrina* haploid yeast stage and dikaryotic multinucleate mycelium with paired nuclei is present. While in *Protomyces* haploid and diploid yeast stage and diploid mycelium is present. In *Schizosaccharomyces* and *Pneumocystis* haploid cells are present. In *Taphrina* septa have simple septal pores and worming bodies are absent. Presence of starch, carotenoid, pigments, sterols, coenzyme or systems are the potential characters for the group.

An asexual soil living yeast, *Saitoella* and *Schizosaccharromyces* are saprobic. Two other genera, *Taphrina* and *Protomyces* are dimorphic, each with a saprobic yeast stage and a parasitic mycelial stage on plant hosts. *Pneumocystis carinni* is the fifth member of this group that is the infective agent of a *Pneumonia*.

**Order Taphrinales**

**Peach leaf curl:** One of the most spectacular disease of this type occurs on Peach and is caused by the fungus, *Taphrina deformans*. Leaves show symptoms at or soon after emergence. Either a part, or the entire leaf blade thickens and curls, the chlorophyll soon disappears and red or purple tints develop within the affected area. Later this appears to be covered by a greyish bloom due to the sporulation of the fungus. Leaves drop prematurely, often so much so that dormant buds are stimulated and further new leaves emerge. These in turn may also infected but not frequently. Repeated losses of leaves in successive seasons greatly reduce the vigour of the tree and its cropping capacity. Blossoms and young fruits can be attacked and these normally drop too but occasionally some infected fruit matures and on these there are prominent warty out growths with reddish tints (Fig. 10.1).

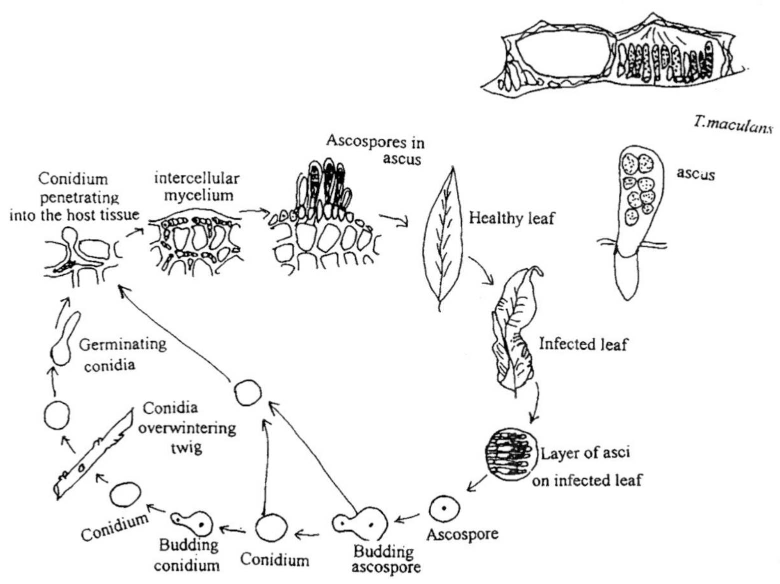


**Fig. 10.1:** Desease cycle of Peach leaf crul by *Taphrina deformans* sp.

Fungus is not an obligate parasite, but generally forms a mycelium only in the host; that may be subcuticular, intercellular and sometimes grow in the cells of the epidermal cells in culture it usually produces cells by budding. It is generally considered to be primitive Ascomycete with affinities to certain yeasts. Mycelium is dikaryotic with multinucleate hyphal compartment. Asci can be clearly seen in sections of diseased leaves. They are not enclosed but form an exposed layer on the leaf surface. These are initially eight ascospores but these produce secondary spores by budding so that many spores may eventually be formed from one ascus. Ascospores and bud conidia can produce secondary spores whenever suitable conditions prevail so that a considerable inoculum is built up. During unfavourable conditions such as hot, dry summers or very cold winters, conidia may become thick walled they may remain viable for two years or more.

Environmental conditions also markedly affect disease severity. Rain is essential for infection, disease is most severe where there is cold.

*T. pruni* – Plum pockets

*T. maculans* – Turmeric

*T. cerasi* – Witches brooms of Cherry

**Protomycetaceae**

***Protomyces*** – Stem Gall of coriander.

This is a common and wide spread disease of coriander causing heavy damage, when associated with wilt.

**Symptoms:** The disease appears in the form of tumour like swellings on leaf veins, leaf stalks, peduncles, stems and fruits. The swelling on the veins give a hanging appearance to the leaves the tumours are at first glossy but later ruptures and become rough. They are about 5 mm in diameter.

The disease is caused by *Protomyces macrosporus*. The mycelium of the fungus is found only in the tumours. Infection is caused by the resting spores of the fungus. The infection become systemic. The hyphae are intercellular and closely septate. Branching is irregular. Scattered cells in the hyphae swell to form ellipsoidal or globose bodies which later develop into chlamydospores. Mature chlamydospores with smooth three layered walls, have a diameter of 50–60 microns. The chlamydospores function as the resting spores of the fungus. Resting spores germinate in water by rupturing the outer wall (exospore). The spore sac is formed by the emergence of cell membrane through the ruptured outer wall of the spore. The protoplasm from the spore passes into this vesicle and gathers towards the periphery. The nucleus divides several times forming 100–200 daughter nuclei. The multinucleate protplast of the spore sac becomes oriented in a peripheral layer, while the nuclei apparently undergo reduction division, resulting in the production of four endospores per nucleus. On maturity these spores separate and collect in the centre of the vesicle (Fig. 10.1 B). The latter burst and the spores are set free. These are capable of multiplying further by budding in a yeast like fashion and cause infection of the host.

Galls are produced as a result of hypertrophy and hyperplasia of the host cells. The galled tissue possess stele derived from the stellar region. In the flower the infected parts become massive. Pollen grains are massed together and become unhealthy. In the infected host parts, the oil canals are severely affected and collapse. The fungs does not invade the xylem.

The pathogen survives in the soil in the form of chlamydospores and this seems to act as the primary source of infection. High soil moisture and shade predispose the plants to infection. Minimum infection occurs at pH 4.6 and maximum at pH 7.4. Loss of sugar, fats, proteins, carbohydrates has been reported from infected fruits.

**Yeast**

The yeast are ubiquitous, unicellular, saprophytic organisms occurring in air and soil and growing wherever there is a sugary substratum available. They have great economic significance and have been in some form or the other connected with the human civilization even since its beginning.

**History:** The first valid description of yeast was done by Leeuwenhoek in 1680. In 1799, Fabroni traced the comparison of yeast cells with the albuminoids. By 1825, it was shown that the yeast cells present in beer and wine increased by the process of budding and in 1839 Schwann for the first time demonstrated the presence of endospores in yeasts. Twenty years later (1859) Louis Pasteur established that the fermentation was distinctly correlated with the life of yeasts. He also introduced pure culture methods which were followed and improved by Hansen towards the late nineteenth century. These techniques greatly facilitated in having a deeper insight into the morphology, cytology and physiology of yeasts.

**Habit and occurrence:** Yeasts usually are saprophytic occurring wherever there is a sugary substratum, in the nectar of flowers, on the surface of the fruits in the milk, animal excreta and also on the vegetative plant parts. Their chief characteristic is to ferment the carbohydrates on which they occur so profusely (Fig. 10.2). They are thus very important occur so profusely. They are thus very important industrially in the brewery and the bakery. They bring about alcoholic fermentation of sugary media in which the resulting products are alcohol and anaerobic forms. Some yeasts occur as parasites also. *e.q. Monosporella* is parasitic on the intestines of *Daphnia* a crustacean; species of *Nematospora* have been isolated from tomato, beans etc. Some forms of yeasts occur in symbiotic relationship with certain bacteria and moulds e.g. the fermentation of rice for the preparation of sake wine in Japan is brought about the *Saccharomyces sake* in association with *Aspergillus oryzae*. The fermentation in the bread is also due to this relationship between bacteria and yeast symbiotic relationship between two parasites *i.e. Oospora linguales* a mould and *Cryptococcus linguae-pilosae* an yeast result to produce the rare infection and disease in man known as black tongue.

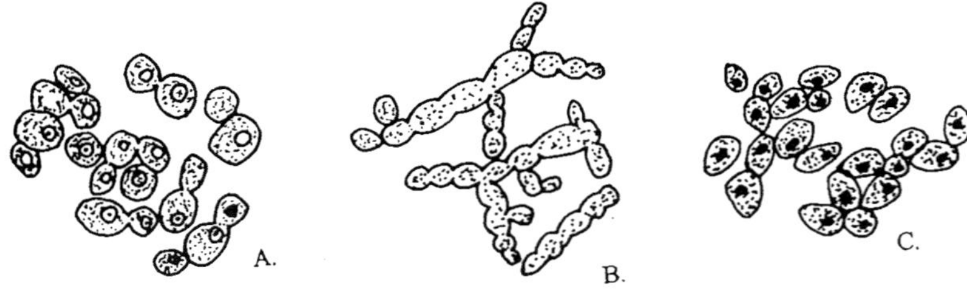

**Fig. 10.2:** *Saccharomyces cerevisiae*

A. budding cell. B. filamentous form. C. & D. Distillary type ascospores

**Vegetative structure:** Yeasts are unicellular organism occurring chiefly in isolated states but sometimes forming unstable pseudo-hyphal structures especially when grown on gelatin–*Schizosaccharomyces. pombe* and *Saccharomyces ludwigii*. Quite often, the daughter buds and daughter cells resulting respectively from the budding and the fission of the parent cell, instead of becoming separated, remain attached to it, thus forming false colonies. The united cells have no solid union and are separated by well marked walls of their own. The individual cells are polymorphic showing different shapes even in the same culture depending upon the nutrition available in the substratum. Generally they are spherical, oval, elliptical, or even elongated usually colourless, but sometimes red, brown, grey or yellow pigments may be observed. The cells are minute measuring about 1–9μ long and 1–5μ broad (Fig. 10.3).

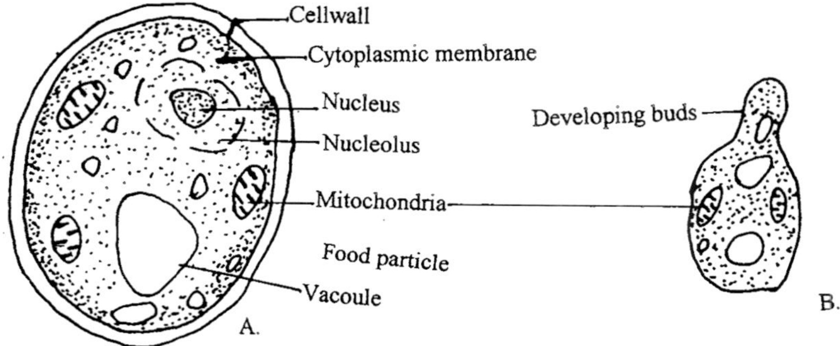


**Fig. 10.3:** Vegetative structure and budding cell in yeast.
A. vegetative cell. B. Developing buds in yeast

A definite two layer cell wall surrounds the cell. It is composed of some chitinous substances beside other compounds. The cell wall is closely followed by a cytoplasmic membrane. The nucleus occupies a variable position with in the cell, is spherical to ovoid and is surrounded by a double membrane which indicates the presence of pores. A distinct nucleolus also occurs at variable positions with in the nucleus. It is a dense more or less spherical structure surrounded by a membrane of its own. It also encloses with it, certain vacoule like areas of denser density. The nucleoplasm is granular and homogeneous. Some irregular and randomly distributed areas occur in the nucleus. These contain fibrils and correspond to the chromosomes of yeast. Besides the nucleus there is a distinct vacoule also present filled with a granular material the volutin and is separated from other constituents of the cell by a membrane of its own. The vacoule is not an integral part of the nuclear apparatus. It is a seprate structure altogether and is not a permanent cell inclusion being absent in the actively dividing cells.

Other cytoplasmic inclusion are the mitochondria, which number 4–20 in a cell endoplasmic reticulum and ribosomes. Besides these, there are other inclusions which are the reserve food plastids of lipids, particles and glycogen etc.

**Asexual Reproduction:** Asexual reproduction takes place by two methods by budding and by fission (Fig. 10.3 B). The yeasts are designated as budding yeast and fission yeasts respectively. Sometimes yeast cells become surrounded by thick resistant walls and their contents become rich in fats and glycogen. These are called durable cells meant for perennating unfavourable circumstances.

**Budding:** Bud initiates at one end of the yeast cell. It enlarges and receives the mitochondria which migrate from the parent mother cell much before the nucleus. Meanwhile, the nucleus divides by a simple process of elongation and constriction which appears in its middle region where from the daughter nuclei separate. One goes into the bud and other remains in the mother cell (Fig. 10.4 A-C). The nuclear membrane remains persistent throughout the division process. So is the case with cytoplasmic membranes remaining closely adherent to the cell wall during the process of cell division. The vacoule may be absent in the actively dividing cells. The bud consequently becomes separated from the mother cell by a constriction at the base and later by a wall. This bud while still attached to the parent, may produce a bud over it or the parent may develop another bud at some other end and thus a long chain of cells is developed sometimes forming branched structures as in *Saccharomyces cerevisiae* the toddy (beer) yeast.

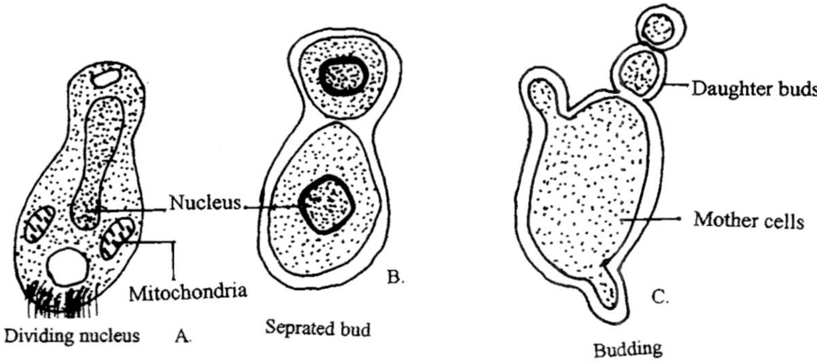


**Fig. 10.4:** Different stages during the budding in yeast. A. dividing nucleus B. seprated bud. C. budding.

**Fission:** Yeast cells under go fission by transverse division the parent cell elongates and the nucleus divides into two daughter nuclei. Simultaneously, a transverse septum develops as annular growth centripetally *i.e.* from the wall towards middle, thus separating the two daughter uninucleate cells. The new wall thickens before the cells separate apart. In most cases, they formed cells instead of separating apart, divide similarly and a false unstable hyphal structure may develop as in *Schizosaccharomyces octosporus*.

**Sexual Reproduction:** Sexual reproduction in yeast is quite varied occurring either between two somatic cells which may be similar or dissimilar or between two ascospores. They are chiefly three types of sexual cycle as described below:–

(*i*) In *Schizosaccharomyces octosporus*, the diploid stage is very short and most of the life cycle is passed in the haploid (x) stage. The vegetative cells are elongate, uninucleate and haploid. Two such cells come close, send out copulation extensions which meet and the separating wall dissolves. The nuclei and cytoplasmic contents migrates into the copulation tube where the two nuclei fuse to form the diploid stage. The fusion occurs by the nuclear membrane dissolving at the site of contact by remaining intact elsewhere. Meanwhile the conjugation bridge widens and the whole structure resembles a dumb bell. The diploid nucleus soon undergoes three divisions, one of which is meiotic. As a consequence, 8 haploid nuclei are formed around which a similar number of ascospores are delimited. Thus the parent cells are directly transformed into the ascus. The ascospores are liberated by the rupture of the ascus wall each of which enlarges and forms daughter cells by fission which continues in the haploid stage. This type of life cycle is designated as haplobiontic.

(*ii*) In the second type, it occurs in *Saccharomycodes ludwigii* the diploid stage is much prolonged while the haploid phase is short. This develops 4 ascospores in each ascus – 2 at each pole. Copulation occurs between two ascospores usually, lying side by side or between those belonging to opposite poles or even between ascospores in different asci. The plasmogamy and karyogamy occurs between the mating ascospores with in the ascus and a diploid cell is formed which pushes through the ascus wall producing a germ tube from which successive yeast cells are formed by budding. This somatic stage is diploid, continuing for long and terminating only when such cells convert into a 4 spored ascus again by the meiotic division of its nucleus.

Thus the ascospore represent the short haplophase. This type of life cycle is suggested as diplobiontic.

(*iii*) In the third type as it occurs in *Saccharomyces cerevisiae* both the haploid and diploid are equally important. To start with two haploid vegetative cells may copulate and form a diploid cell which continues growth for long by budding. The diploid nucleus in these cells at some time undergoes reduction division and 4 haploid ascospores are produced when these are set free from the ascus, they too sprout profusely by budding but are smaller than the diploid cells. Thus the diploid and haploid cells occur commonly in the vegetative phase. This type of life cycle is designated as haplo-diplobiontic (Fig. 10.5 A-I).

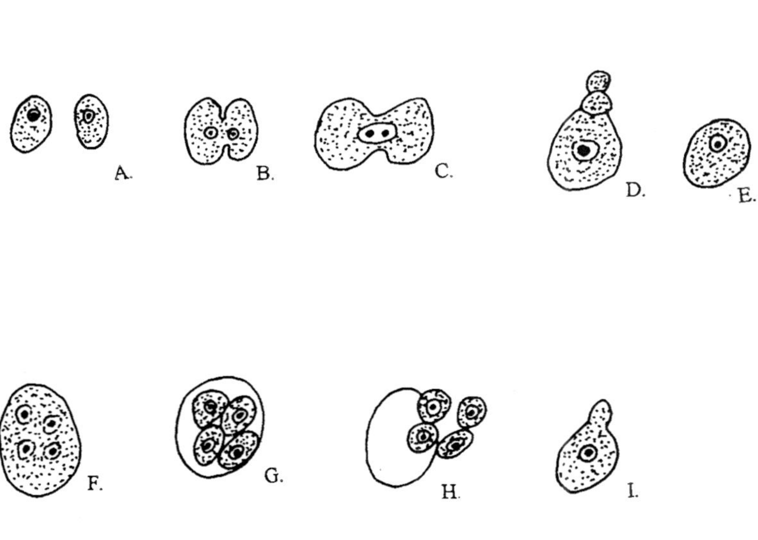


**Fig. 10.5:** Sexual reproduction in *Saccharomyces cereviciae*

A, B & C. haploid copulation cells, D & E. Diploid cell. F. Dividing fusion nucleus. G & H. Asci with Ascospores. I. Haploid cell (ascospore).

There is parthenogamy *i.e.* formation of asci without actually involving couplation of two cells *i.e.* in species of

*Zygosaccharomyces*, *Saccharomyces* and *Schizosaccharomyces*. There are also species which donot form asci at all as *Schizosaccharomyces asporus*.

Three facts appear evident in the sexuality of yeasts. Firstly, that there is degenerate sexuality as exhibited by copulation between vegetative cells; Secondly, that there is irregular and uncertain sexuality as exhibited by fusion between morphologically similar sprout cells and also between ascospores and thirdly, that the ascus necessarily develop as a result of copulation. Its formation being dependant upon nutritional conditions also.

Number of ascospores is variable. It may be 1 in (*Monospora* and *Nadsonia*) 2–4 in (*Debaryomyces* and *Hansenula*), 4 in *Saccharomyces* and 8 in *Schizosaccharomyces* (Fig. 10.6. A-D). A several ascus occurs in *Kluyveromyces polysporus* discovered by Vander walt in 1956 in South Africa.

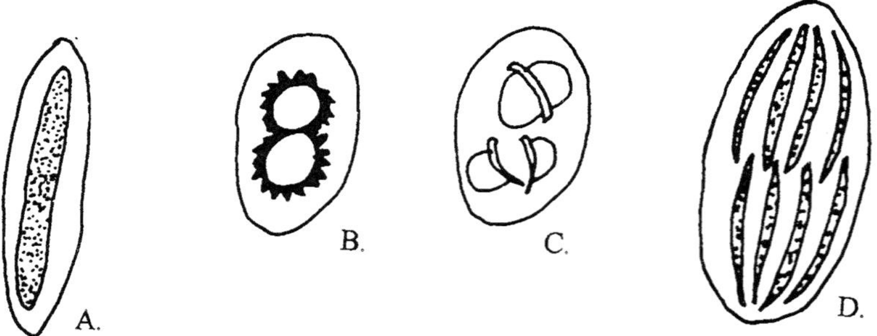


**Fig. 10.6:** Different forms of ascospores in yeast.

A. *Monospora* B. *Debaryomyces* C. *Hansenula*. D. *Nematospora*.

**Plectomycetes:** The filamentous ascomycetes develop functional sex organs, but sexual degeneration especially of the male gametangium, has demonstrated in a number of species and appears to be trend in some group. Most ascogonium, ascogenous hyphae and croziers that come to be enclosed in an ascocarp–a cleistothecium, perithecium, apothecium or ascostroma. Plectomycetes have the various characters (Fig. 10.7 A-D).

(1) Asci are thin walled, globose to pyriform.

(2) Asci scattered at various levels within the ascocarp not forming a hymenium, arising from ascogenous hyphae of various length ramifying through the ascocarp.

(3) Ascospores unicellular.

(4) Ascocarp typically a cleistothecium, when one is present.

(5) Cleistothecial peridium varying from thin wefts of hyphae forming an arachoid covering over the asci to reticulate cage like arrangement of hyphae.

(6) Various types of anamorphs and conidial forms, characteristics of certain families and orders (Fig 10.7).

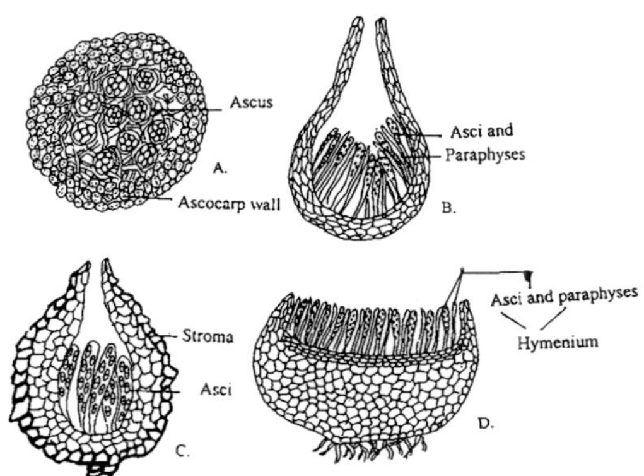


**Fig. 10.7:** Various types of fruiting bodies

A. Celistothecia B. Perithecium C. Ascostroma D. Apothecium

***Aspergillus:*** It is one of the common saprophytic moulds occurring on decomposing organic substance e.g. bread, damp fruit and vegetables, and cheese etc. At first the mould is white, but later it assumes a greenish colour, hence its popular name "Green mould". This same fungus is known to different botanist under the names of *Aspergillus glaucus* and *A. herbariorum* (Fig. 10.8).

Some species of *Aspergillus i.e. A. flavus* and *A. terreus* causes fungal diseases in animals and human beings known as **Otomycosis**. *Aspergillus fumigatus* causes infection in the lungus and symptoms appear similar to tuberculosis. Some species of *Aspergillus* have been found to possess antifungal and antibiotic properties. *A. terreus* yields terrein and *A. fumigatus* fumigatin.

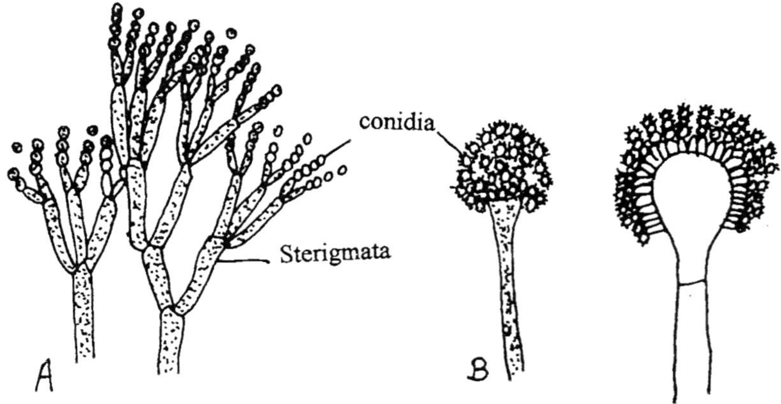


**Fig. 10.8:** Conidial heads of *Penicillium* and *Aspergillus*

**Mycelium:** The vegetative growth consists of a tangled mass of hyphae on the surface and finer feeding hyphae penetrating the substratum. The hyphae are septate, even in the youngest parts, thus differing from previous types, each segment contains granular protoplasm, several nuclei and oil globules.

**Asexual Reproduction:** From the aerial mycelium arise numerous straight, non septate, erect branches the conidiophores. Conidiophores enlarge and develop a thick wall. This is the foot cell from which a vertical branch, the conidiophore develops perpendicular to its long axis. The conidiophore elongates and swells up apically into a vesicle. Large number of nuclei and the protoplast migrates into the vesicle from the parent hypha through the conidiophore. The conidiophores generally, are unbranched, aseptate, thin and smooth walled, outer wall may be pitted or irregularly thickened. Each coindiophore forms a spherical head, but this is not cut off by a septum, all over this head arise peg like outgrowths, termed sterigmata, and form the apex of each sterigmata as it elongates conidia are abstracted. The sterigmata are mostly placed perpendicular to the surface of the vesicle. They may be hyaline or pigmented, and formed in one series (uniseriate) or in two series (biseriate). The secondary sterigmata develops over the first one primary sterigmata. Often the primary sterigmata become more elongate than the secondary series (*A. niger* and *A. candidus*) Two series of sterigmata are also equal in some species (*A. nidulans* and *A. versicolor*).

One or more nuclei are received in the sterigmata from the many which are present in the vesicle and have earlier been received from the parent hyphae. When the sterigmata are in two series, the second series receives similarly one or more nuclei resulting from the nuclear division in the primary sterigmata. A septum seprates the conidium from the sterigmata. Simultaneously, another one is formed by the sterigmata below the first one. The successive production of conidia, in long chains from the tip of one sterigmata may number upto 100, oldest at the top and the youngest still forming over the tip of the sterigmata. Each of these conidia receives nuclei from the mother sterigmata by successive division. The conidia, until they are fully matured into their characteristic size and shape, draw nourishmet from the parent cell. After septation it secrets a fresh wall of its own; seprates from the primary wall of the sterigmatas. The youngest conidium is next the sterigmata and successively older ones form a chain with the oldest at the outer free end. The mature conidia are readily shed or carried off in the air, and their places taken by the next ones below. These small oval conidia are green in colour, but the pigments occurs in the wall of the spore, and is not chlorophyll, therefore plays no part in carbon assimilation. It is owing to the production of enormous numbers of these conidia that the white mycelium assumes the better known green colour, and the fungus receives its popular name. Conidia are always present in the atmosphere, and on reaching a suitable substratum they germinate readily, emitting germ tubes which develop to mycelia (Fig. 10.9).

**Sexual Reproduction:** The same mycelium which has produced conidia eventually bears sexual organs. The general features of these organs were first described by de Bary in his classical researches of 1870.

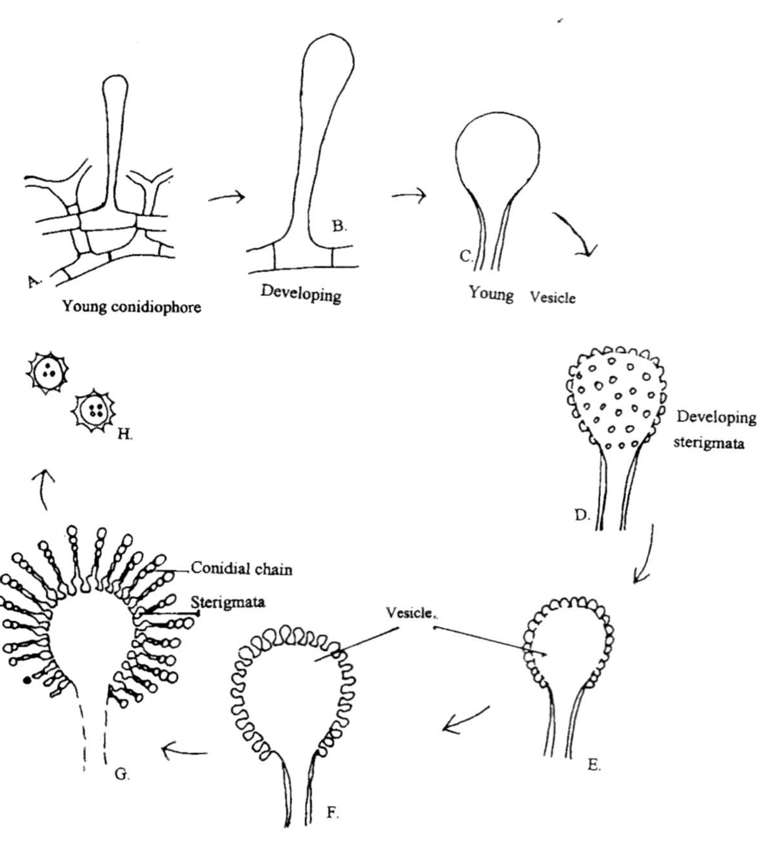


**Fig. 10.9:** Asexual reproduction in *Aspergillus*.

The female organs the archicarp, is an erect hypha of which the upper part is coiled like a cork screw. The coils are at first loose and open, but later becomes closely compacted. The whole organ consists of a multicellular stalk, with a single apical cell, the ascogonium, corresponding to the oogonium of *Pythium* (Fig. 10.10).

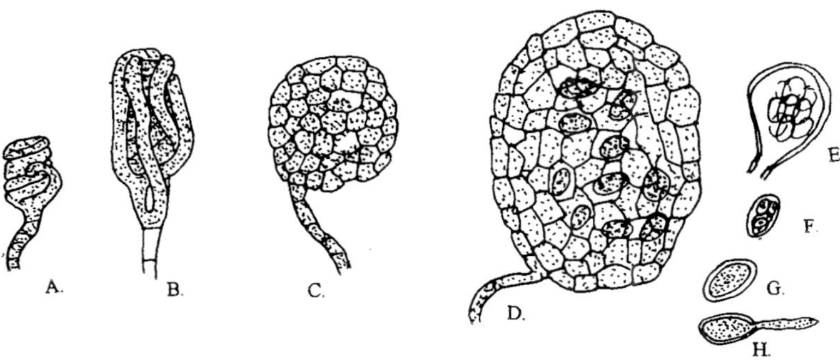


**Fig. 10.10:** Sexual reproduction in *Aspergillus*

A-D. formation of cleistothecium. E. mature ascus. F. ascospore. G-H. germinating ascospore.

Near the archicarp another septate hyphae appears from the end of which is cut off by a septum a single cell, which is the male organ, or antheridium. The antheridium arches over until tip comes into contact the intervening walls break down, and the process of fertilization is effected by fusion of the contents of the male and female cells. After fertilisation the ascogonium becomes septate, and from its segments branches grow out, the nuclei pass into them, and the ends are cut off by septa. These function as sporangia, and with in each of them eight spores are developed. On account of their very distinct mode of formation these sporangia are distinguished as asci and their spores as ascospores.

Before the ascogonium becomes septate, much branched, sterile hyphae, arising from the stalks of the sexual organs, begin to grow up about the latter, by intertwining and septation. These hyphae form a mass of pseudo-parenchyma in which the ascogenous branches become embedded, the outer layer of this structure forms a protective sheath or case, while the inner filling tissue is used up for the nourishment of the developing spores. As this frutification has no opening, it is called a cleistothecium. Cleistothecium is small, globose, smooth walled and is generally golden yellow in colour. When mature the asci are disorganised, and the ascospores lie free with in the ascocarp, to be liberated in due season by the rupturing of the wall of the later. These golden yellow ascocarps are of a resting nature and serve to carry the fungus over a period unfavourable of growth. By the time ascospores are mature. They are notched at each end and then liberated by suitable conditions of moisture and temperament.

**Economic importance**

1. Species of *Aspergillus* causes lung disease i.e. *Aspergillus fumigatus*.
2. *Aspergillus flavus* and *A. terreus* causes disease in ear known as otomycosis.
3. Many species of *Aspergillus i.e. A niger, A. flavus, A repens* etc. are responsible for the spoilage of exposed food stuffs, jams, jellies, pickles and other sugary substance.

**Useful activities**

*Aspergillus oryzae* is used in preparation of wine.

*A. niger* can detect copper even in traces.

*Aspergillus* used in cheese manufacture.

*A. gossypii* used in the manufacture of vitamin B.

***Penicillium:*** Mainly saprophytic, it occurs on decaying vegetables, fruits, meats, soils and in forest floors on fallen and rotten plant remains. *Penicillium expansum* causing rotting and spoils fruits like apples, bears, grapes etc. in storage. *Penicillium purpurogenum* occurs commonly as spots on printing papers, books and engraving. Textile fibres, paper pulp and lumber woods are discolourised and made inferior by the saprophytic growth of *Penicillium* over them. While *P. notatum* and *P. chrysogenum* have antibiotic properties. These have been used to yield to most successful antibiotic drug 'Penicillin' (Fig. 10.11 A-B).

**Vegetative structure:** Mycelium develops as ramification of profusely branched and septate hyphae either penetrating deep into the substratum or forming a superficial weft over it. Cells are thin walled and contain 2 to several nuclei in them. The initial branches grow in a radiating manner and their interspaces are quickly filled by further branching system. The colony presents a velvety floccose or funiculose appearance the hyphae of most species are variously coloured due to the production of pigments on the surface of the hyphal walls. Many species exude 'transpiration fluid' drops over the colony which have characteristic colour and odour. Sterile areas or over growth and sclerotic also develop in some species *i.e. P. italicum* and *P. gladioli*.

**Asexual Reproduction:** The conidiophores arise from aerial hyphae as well as from those that are submerged in the substratum. They lack differentiated and a distinct foot cell, walls are smooth sometimes roughened due to granulations, echinulations or single or may form clusters fasicle or definite coremia. These may be one of the following two distinct types: (Fig. 10.11).

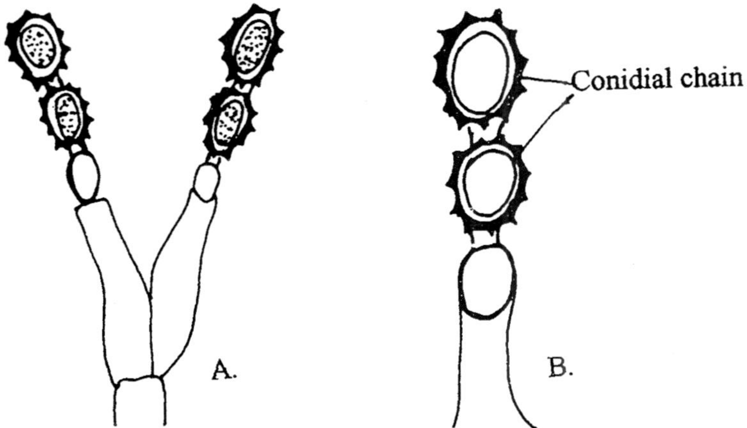


**Fig. 10.11:** *Penicellium*—Conidial development
A. Biverticillate B. Monoverticillate

1. **Monoverticillate:** When the conidiophores are produced as septate, single structures and directly bear a terminal cluster of sterigmata and conidial chain (Fig. 10.11 B).

2. **Biverticillate:** When branching in the conidiophore occurs at two or rarely more than three levels below sterigmata. They may be of two types:

(a) *Asymmetrica:* When the entire conidial head appear one sided *P. expansum.*

(b) *Symmetrica:* If the branching is regular and symmetrical on both sides of conidiophore, so that the conidial head appears uniform and even *i.e. P. variable*. Very few species possess such conidiophores which branch several times below the level of sterigmata. Second set of branches is known as metulae. Hyphae, conidiophore and sterigmata are multinucleate. Number of sterigmata in a whorl may be only few or very much more. These terminally bear a large number of conidia in long chains. The conidia are uninucleate, sometime binucleate.

Conidia are formed by the segmentation of sterigmata. Conidium segment lengthens and swells to attain the shape and size, characteristic of the species which may be globose or elliptical. When mature it secrets a wall within the primary wall. Both may remain distinct or coherent to appear as one. Outer wall of a mature conidium may be smooth or roughened being, echinulate or sculptured. It is comparatively thick, sometimes reaching 1/5th of its diameter. This conidium is pushed up by the next one formed similarly at the apex of the sterigmata; thus a chain of conidia develops in which the oldest is at the top and youngest remain attached to the sterigmata.

In *P. digitatum, P. tardum* conidia are formed inside the sterigmata tube, from which conidia slips out. In some species the vestigeal cell wall between the two conidia, in the chain remain distinct to form a connective conidia are disseminated by the air and falling on favourable substratum in presence of suitable moisture and temperature, germinate by one or more germ tubes to form the new mycelium.

**Sexual Reproduction:** Multiplication of the fungus mainly takes place by conidia. During the lack of oxygen high sucrose content in the substratum favour development of the ascocarp (Fig. 10.12).

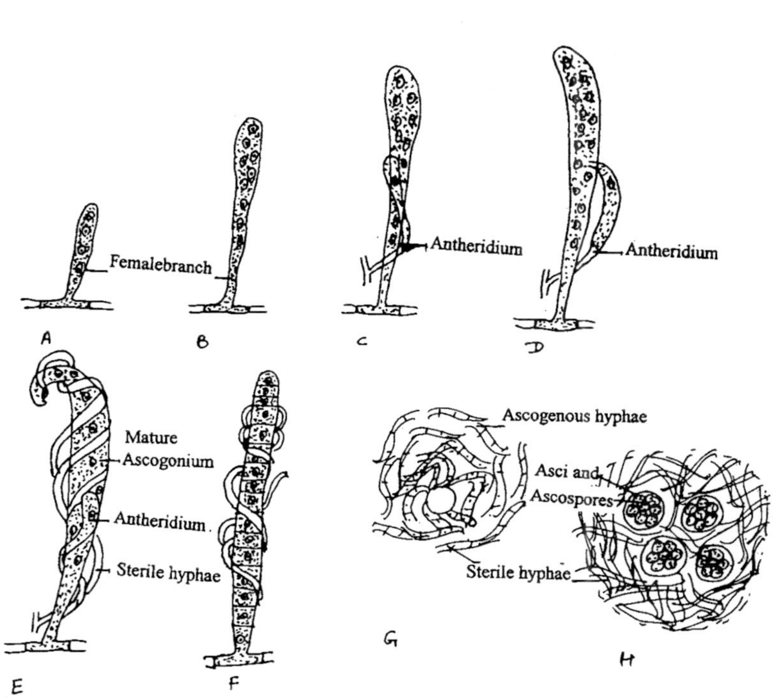


**Fgi. 10.12:** Sexual reproduction in *Penicillium*

Except *P. luteum* all species are homothallic. Sexual reproduction either may be isogamous or anisogamous.

(*i*) In *P. bacillosporum* sexual branches are isogamous. They arise as two short multinucleate branches from cells of the neighbouring hyphae and become helically coiled around one another. The two branches fuse by a large pore, the ascogenous hyphae appear within the ascocarp and give rise to asci arranged end to end to form short curved or coiled chains. Each ascus contains 8 globose ascospores whose walls are marked with coarse spine (Fig. 10.13).

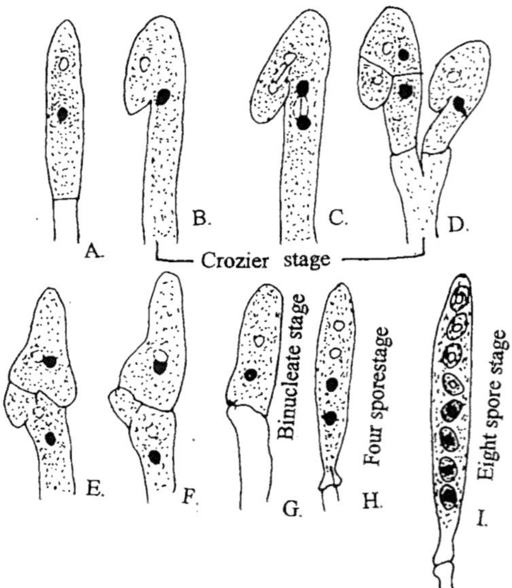


**Fig. 10.13:** Ascus development

(*ii*) In *P. vermiculatum* sexual branches are anisogamous *i.e.* an elongate ascogonium and a short antheridium. An erect and long branch destined to become the ascogonium that develops from the hyphal cell. Young ascogonium is uninucleate and unicellular. As it matures, nucleus divides into 32–64 nuclei. From the another branch antheridium initial develops, close by. This is also uninucleate. Antheridium coils around the ascogonium. From mature antheridial branch, antheridium proper cuts off, which is uninucleate, gets swollen and becomes closely appressed to the separating wall between the two dissolves and after this ascogonium divides into a row of large number of cells each containing two nuclei. Some of these cells develop into one or more branched ascogenous hyphae and the dikaryotic cells, toward their tips, 8 spored asci which may be held in short chains. Simultaneously sterile hyphae grow to form loose or compact tangle of hyphal mass surrounding the sex organs and the asci.

Ascospores may ripen with in 7–10 days, or may require even one month or more to do so. These are variable in shape, may be spherical, elliptical or lenticular smooth walled or the outer walls may be echinulate, pitted or distinctly banded showing a pulley wheel like structure *i.e.* two distinct ridges with a furrow along the equatorial plane. The ascospores falling on a suitable subtrate swell up and germinate by germ tubes to form the new mycelium..

**Penicillin:** Penicillin was discovered by Prof. Alexander Fleming in 1928. He had observed that Penicillin which came up as a contaminant, in a culture colony of *Staphylococcus* did not allow it to grow. He cultured this blue green mould in pure growth and isolated the bacteriostatic substance and named it 'Penicillin'. This substance was later found to be most effective against the gram positive bacteria. The potential therapeutic value of Penicillin was realised only in 1940 by Prof. H.W. Florey, E. Chain and others at Oxford University. Efforts then were made to evolve the most suitable culture media and techniques for obtaining the best growth of *Penicillium* and also to determine and develop more productive strains. Penicillin is 'wonder drug' being most effective against pyogenic cocci and the Gram positive spore forming Bacilli including forms of *Clostridium*.

**Erysiphales:** Fungi belonging to Erysiphales are obligate biotrophs that cause a major group of plant diseases commonly known as powdery mildews. There are so named because the leaves are covered by white powdery material. Their mycelium form an extramatrical superficial layer over the host surface occurring mainly as ectoparasites. In favourable condition of growth, they reproduce asexually by formation of conidia on aerial conidiophores either singly or in chains. The sex organ consists of a single ascogonium and an antheridium. The cleistothecia are ovate to spherical without any opening and are externally furnished with characteristic appendages. The mature asci form a parallel layer on the floor of the ascocarps. Powdery mildew attacks on the dicots mainly. Examples of economically important host include cereal grains, cucurbits, apples and grapes. Some powdery mildew seems to be mild but they are very destructive. These are the fungi that produce unitunicate asci in hymenia, wholly borne as a basal layers in ascocarps. The fungi usually designate as pyrenomycetes, produce their typically club shaped or cylindrical asci in a hymenium that lines the base and often the sides of the inner wall of the ascocarp. The asci are usually persistent, but may be evavescent. The ascocarp is a perithecium in most species. The ascocarp may or may not be embedded in a stroma, but have a definite wall (Peridium).

**Sexual Reproduction:** Sex organs arise as one celled and uninucleate structure the female ascogonium and male antheridium both usually supported one stipe cells. Their tip unite and the male nuclei migrate into the ascogonium cell. The asci are formed in the usual manner which generally do not mature as they are formed but require long intervals and may ripen even next season. The number asci per cleistothecium is generally several but sometimes, only one ascus may be formed in *Sphaerotheca* and *Podosphera.* When several, they are arranged in a parallel manner at the floor of the cleistothecium which when young are colourless, yellowish when immature and brown to black when fully mature. The cleistothecial wall is 2 layered, outer composed of the thick walled cells without nuclei, while the inner layer consists of thin walled hyphae whose cells contain abundant protoplasm. The cleistothecia, along their equatorial plane usually are furnished with appendages which are characteristic of the genera and help them to attach on the host surface before and after dissemination. Mycelium like appendages occur in *Sphaerotheca, Erysiphe* and *Leveillula* appendages dichotomously branched at their tips occur in *Podosphaera* and *Microsphaera*; appendages with coiled tips occur in *Uncinula* and acicular appendages with bulbous base are characteristic of *Phyllactina.* The cleistothecium will bursts irregularly due to the swelling of the asci with in it. The ascospores are set free which germinate on the host surface to form the new mycelium. A characteristic feature of the family is that the mother spore remains a living integral part of the colony or the thallus and does not perish or collapse after the fungus has established nutritive relations with the host.

***Erysiphe:*** *Erysiphe* is a cosmopolitan powdery mildew occurring as an ectoparasite mostly on cultivated plants forming an extra matrical mycelium on the host surface. The species generally have a very wide host range but may also exhibit specialisation of host and are then restricted to a single host species or to a small group of closely related species. *E. polygoni* (Fig. 10.15 A-D) affects as many as 357 species distributed in 157 genera belonging to 42 families, chief of them being poppy, clover, bean cabbage etc. *E. graminis* is wide spread on oats, barley, wheat, rye etc. (Fig. 10.14 A-C).

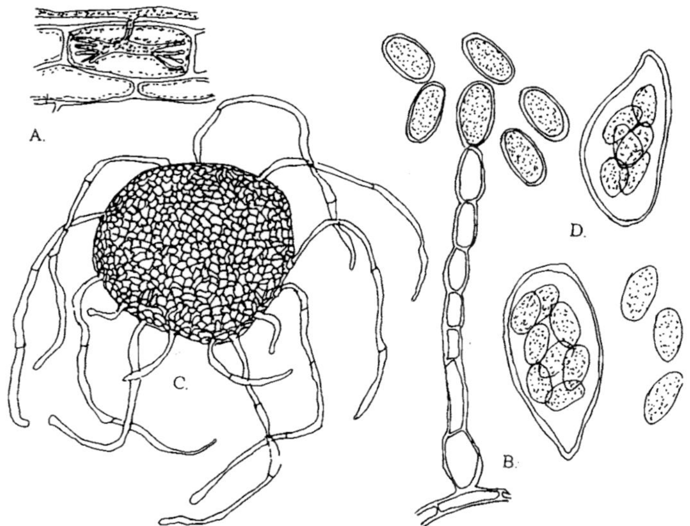


**Fig. 10.14:** *Erysiphe graminis*

A. Digitate haustorium. B. B. Conidia present on conidiophore. C. Clestothecia D. Ascus with ascospore

**Vegetative Structure:** Mycelium is ectophytic spreading along the host surface superficially. Endophytic growth is reported in *E. graminis* if the conidia germinate on the wounded host leaf and *E. taurica* occurs as a hemiendophyte because the mycelium occurs in the tissue and the conidiophores emerge through the stomata. In general mycelium comprises branched hyphae with short uninucleate cells. The hyphae develop simple or lobed appresoria where they are in contact with host epidermis. The epidermal cell wall swells up and the narrow slender hyphal tip bores its way through the cuticle to penetrate into the epidermal cell where it develops into a bulbous or saccate haustorium. In *E. graminis* however, the haustorium is formed as digitately branched tubular processes running parallel within the cell.

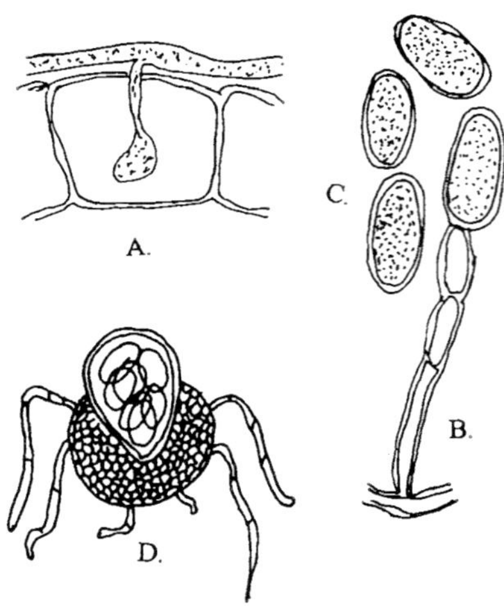


**Fig. 10.15:** *Erysiphe polygoni*

A. Haustoria in cell. B. Conidia present on conidiophore. C. Conidia. D. Cleistothecia.

**Asexual Reproduction:** After the mycelium has established itself on the host surface the hyphae produce short and erect uninucleate conidiophores vertically. The conidiophore may behave in one of the following ways.

1. It may cut off conidia terminally in an acropetal succession as in *E. cichoracearum* these may be held in a chain.
2. It may be divided into a short apical cell and a long basal cell forming the stipe, the former generative or conidium mother cell cuts off conidia successively as in *E. polygoni*. The mother cell is thus separated by a long stipe cell from the sporiferous hyphae.

3. The apical cell of conidiophore divides to form two conidia and similarly the next cell is cut off below it which again forms two conidia. Thus a chain at about 20 or more conidia is formed of which the terminal two conidia are equal and are the oldest pair followed basipetally by similar but younger and younger pairs as in *E. graminis*.

Conidia are hyaline, uninucleate, highly vacuolate thin and smooth walled, ovate sometimes, papillate on one or both ends. Conidia are produced in abundance and are disseminated by wind. The conidia falling on a suitable host surface, germinate immediately in damp air or water by putting forth one or more germ tubes developing from their narrow ends. Within a few days the new mycelium formed from the germinated conidium begins to form a fresh conidial crop and this process continues through out the active phase of growth of the host.

**Sexual Reproduction:** Towards the end of the active period of growth of the host, the sex organ appear at hyphal ends on those branches which lie either coiled or parallel to each other. One of these, the ascogonium is comparatively large and the other the antheridium is slender. Sometimes a distinct antheridium may not be recognisable. Both are uninucleate i.e. tips come in contact, the separating wall dissolves and the male nucleus migrates into the ascogonium. After the plasmogamy, the nuclei divide and a row of 4–5 cells arise by separation in which the penultimate cell is generally dikaryotic or sometimes more nucleate. The ascogenous hyphae develop from this cell and form a dense tangle of branches. The ascus forming cell always contain the dikaryon, these cells enlarge, the two nuclei fuse and the diploid nucleus divide by 3 divisions, forms 8 nuclei which becomes round and the ascospores are delimited. The number of ascospores generally do not ripen until the next season. Meanwhile a hyphal tip usual rises up to surround the ascogonium and forms the wall of cleistothecium. This develops from the cell subtending the ascogonium. To start with, the wall may be single cell thick but gradually, it may become 6–10 cell thick in mature cleistothecia. From the outer layer of thick walled cells which do not contain any protoplast in them, arise elongate unbranched hyphal appendages, which half the cleistothecium to remain fixed up on the host. It may be disseminated by wind and may fall on fresh host. After over wintering the upper portion cracks irregularly, or sometimes wall may split transversely along the equatorial plane. The ascospores may be ejected out violently and germinate immediately if they fall on a suitable host, conditions favouring the conidial crop is produced with in a few days.

***Phyllactinia:*** It is a cosmopolitan powdery mildew parasitising over 100 species of plants, chiefly the deciduous trees. It is a hemiendophytic parasite with profuse extra matrical mycelium on the host surface. *P. corylea* is a very common species parasitising *Corylus avellana* (Fig. 10.17), *P. dalbergiae* (Fig. 10.16) occurs on *Dalbergia sissoo* and is a common powdery mildew.

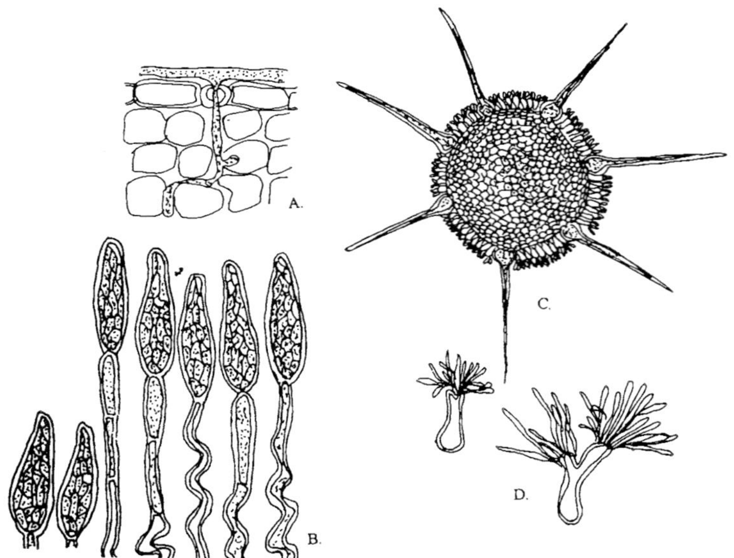


**Fig. 10.16:**

A. Haustoria in intercellular space. B. Oidia formation. C. Cleistothecia. D. Penicilloid appendage of *Phyllactinia dalbergiae*.

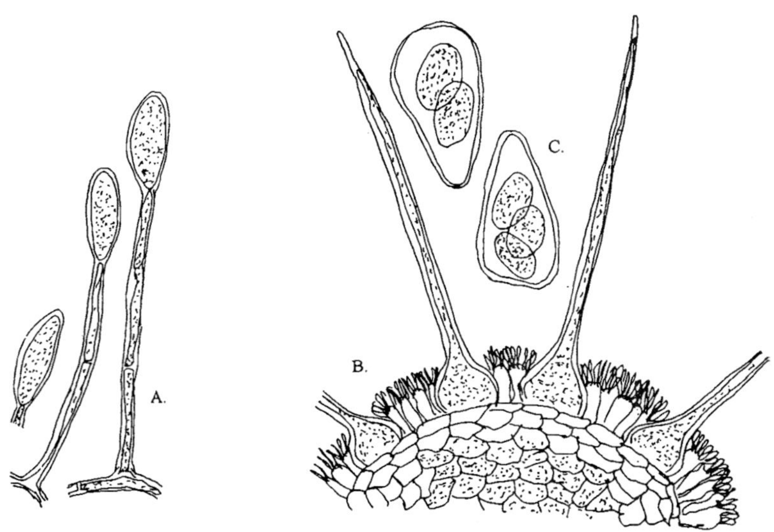


**Fig. 10.17:**

A. Conidia on condiophore. B. Cleistothecia with bulbous appendages. C. Ascus with Ascospore. *P. Corylea*

**Vegetative Structure:** The mycelium is superficially extra matrical over the host surface, mainly affecting the foliage and spreading on both surfaces of the leaves. The hypha are septate with uninucleate cells. Haustoria of *Phyllactinia* are not formed in the epidermal cells from these hyphae but instead a small 5–7 celled hyphal branch of limited growth pierces through the stoma, this develops in the substomatal chamber in the intercellular spaces of the adjoining mesophyll cells. The saccate haustoria are formed from these branches which penetrate the bordering cells.

**Asexual Reproduction:** The vertical conidiophores develop from the superficial mycelium. These are formed in abundance first on both surfaces of the leaves but later become restricted only on the lower surface when the cleistothecial development initiates on their upper surface. The conidiophore divides into the upper generative or the conidium mother cell and a basal stipe cell. The former cell divides again into two, the upper forming the conidium. When this is shed off then the subterminal cell again divides and produces another conidium, similarly conidia are formed singly one at a time, not in chains. They are formed in great numbers till the favourable conditions last. The conidia are clavate, one celled uninucleate, hyaline and thin walled. These are disseminated by wind and soon germinate into a new mycelium which again repeates the cycle within a few days.

**Sexual Reproduction:** Sex organ arise on superficial hyphal branches that lie close together or are coiled. Both are one celled and uninucleate they may cut off a basal cell to support them, which too is uninucleate the nucleus being received by the division of the initial nucleus in the respective sex organs. The ascogonium is thick, ovoid and is closely applied to or is twisted around a small antheridium. The separating wall between the two dissolves and the male nucleus migrates into the ascogonium. Both the nuclei pair and divide the fertilised ascogonial cell divides to form a row of 3–5 cells, of which the penultimate cell is dikaryotic, containing one nucleus from each of the sex organ. The cell produces the ascogenous hyphae from which are formed the asci. Meanwhile, the hyphal branches arise upwards from the basal cell of the ascogonium to form a 2–3 layered sheath around the developing ascocarp. Internally the dikaryon, with in the young ascus, fuses. The diploid nucleus divides by three division into 8 nuclei, first division is reduction division. The ascospores are formed round only two nuclei, the root disorganise. The mature ascus therefore is only 2–spored. Large number of clavate asci form a parallel layer on the floor of the ascocarp and may not mature until the next season.

From the outer most layer of cleistothecium arise characteristic appendages round its equatorial plane. These are long, unbranched, setiform, rigid structures with a bulbous or saccate base. The upper portion of this sac is thick walled and its lower side is thin walled. In dry weather the lower side becomes flaccid, bends down like a joint and the cleistothecium thus is raised up as if on slits. It can now be disseminated by wind. The appendages thus perform the propagative function. Besides the appendages there is a crown of penicillately branched hyphal cells on the cleistothecial top. They are hygroscopic and swell upto form an apical drop of slimy substance, which when dries up, fixes the cleistothecium firmly on its substratum to the host. These cells thus perform the clinging or the attaching function. Cleistothecial wall ruptures irregularly to escape out the asci from which the ascospores are liberated. The ascospores are ovate to elliptical or even spherical and are uninucleate. On a suitable host surface, they germinate to form the new mycelium which is soon ready to produce the conidial crop. In *P. corylea* parasitic on *Corylus americana* there is normal fertilization between the sex organs and the male nucleus is transferred into the ascogonium while in *P. corylea* occurring on *Corylus avellava* there is no fertilization and development is apogamous.

***Uncinula nectator:*** causes powdery mildew of grape vine. *Sphaerotheca* causes powdery mildew of goose berries.

**Vegetative Structure:** Superficial mycelium is hyaline and on account of the copious production of conidia appears as if flour or white powder has been dusted on the plant parts. This gives the general name of powdery mildew to this group of fungi. The hyphae develop radiating from the point of infection and grow closely appressed to the host surface. The cells of hyphae are thin walled uninucleate and vacuolate. Lateral swellings termed appresoria develop along the hyphae, usually one for each alternate hyphal cell by which they remain attached to the host surface and initiate haustorial formation the haustoria generally are saceate or bulbous although digitately branched structure occur in *Erysiphe*. They generally penetrate only the epidermal cells except a few forms in which they may be formed with in the sub-epidermal cell as in *Uncinula* (Fig. 10.18 A-C), in the mesophyll cells bordering the sub stomatal chamber as in *Phyllactina* or in the mesophyll tissues as in *Leveillula*.

**Asexual Reproduction:** It occurs by means of conidia formed apically over conidiophores produced at right angles to the host surface. Conidia are borne either singly or in chains. The abstriction, formation and maturation of the conidia occurs mainly during the light period of the day while the lengthening of the generative cell and the increase in the conidium size and structure takes place during the night. The conidia are uninucleate and highly vacuolate. They germinate by germ tube and reinfect the fresh host parts. The conidial life cycle may complete in about 5 days. The conidia are turgid due to high water content in them and that is why these fungi have luxuriant growth even in dry weather.

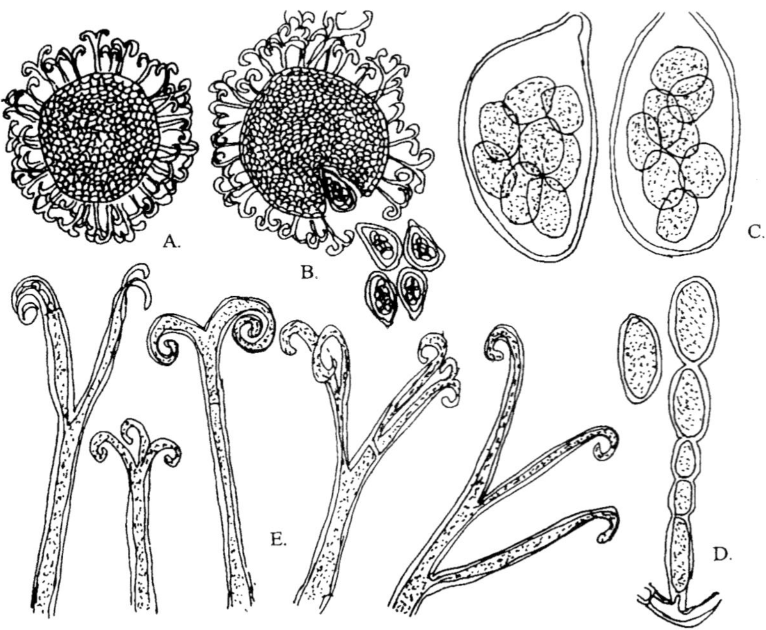


**Fig. 10.18:** *Uncinula nectator*

***Sphaerotheca:*** American Gooseberry Mildew–This disease was probably introduced into Europe from U.S.A. by the introduction of diseased stock. It was first recorded in Ireland in 1900, and since then has spread into many other countries.

**Symptoms:** The fungus occurs on the leaves, shoots, and fruits of gooseberry bushes. In its earliest stage it appears as a delicate web on the expanding leaves but it is more easily detected a little later when the growth becomes white and powdery, as if the affected parts are dusted with flour. Later these patches turn brown, run together irregularly, and form larger areas with the threads woven together like felt. On this brown felt numerous small black dots can be seen with the naked eye. Similar patches occur on the young wood, and the tips of affected shoots look brown and shrivelled. The fruit first attacked, and most seriously injuried is usually that on the shaded parts of the bushes (Fig. 10.19 H-K).

**Fungus:** The web like growth on the leaves and shoots is the epithetic mycelium. The hyphae in contact with the epidermis haustoria are formed, which pierce the epidermal cells, with in which they expand into bladder like form; they serve for both absorptive and fixative purposes. Being thus robbed of nutrient, the leaves and shoots are checked in growth and the bushes seriously injured. The berries are generally attacked at one side and owing to the retarded growth these, become curved on deformed; this part often cracks and decays.

The white powdery appearance of the mycelium is termed the summer stage, and is due to the innumberable conidia borne on short, erect conidiophores arising from the mycelium. The conidiophore is divided by transverse septa into a chain of uninucleate cells, each of which becomes rounded at the ends as it mature, the oldest being at the apex. These are the conidia, which are dispersed as they mature throughout the season. Each conidium that reaches a delicate part of a bush may germinate, cause infection, and become a new centre of disease.

As season advances the mycelium becomes dingy brown in colour and bears perithecia just visible to the naked eye as black dots. Each perithecium is produced by a sexual process between as ascogonium and an antheridium and consists of a spherical or egg shaped wall enclosing a single ascus; from the exterior of the perithecium, and near its base, grow out a number of simple, unbranched, hyphae, which at first serve to fix the structure to the mycelium, and later aid in the dispersal by wind or insect. The perithecia fall to the ground before or with the withered leaves, and may lie there until spring. They are generally considered as 'resting spores' they may be found during summer quite mature and capable of germinating with in a few hours. Such spores may cause infection of new growth after autumn pruning, when conidia are absent. The ascus itself is a membranous, oval sac and contains eight oval ascospores. In the process of germination the perithecium is ruptured at the apex, the ascus protrudes and is ruptured at its free end and the ascospores may be ejected an inch or more. The requisite force is produced by the absorption of moisture by the ascus. The winter spores in soil or debris or on the bushes, germinate during May and June and start the seasonal out break. No part of the fungus persists through winter in the tissues of the plant.

*S. humuli* Strawberry mildew

*S. pannosa* Rose mildew

***Podosphaera***

***P. leucotricha:*** Apple mildew–This mildew is one of the commonest pests of apple, trees and causes serious damage to both old and young plants. It covers the leaves and tips of the shoots with a dense mycelial growth. The powdery stage being very conspicuous quite early in the season. Perithecia are comparatively rare, and the fungus probably carries over winter, mainly by mycelium which lies dormant between bud scales and in other cervices or actually with in the buds.

In *Podosphaera* perithecium bears appendages at the apex or the equatorial region and tip of these appendages are forked dichotomously.

*P. oxycantha* Cherry mildew (Fig. 10.19 A-C).

***Microsphaera*** and ***Uncinula*** (Fig. 10.19 D-G).

*M. grossulariae:* European Gooseberry Mildew. This mildew is much less injurious than *S. morsuvae*. Mycelium occurring on the leaves remains thin and scanty throughout the season, never becoming felted, it rarely spreads on to the shoots. Perithecia are exposed, not embedded in the mycelium, and their appendages are branched at the ends somewhat like the fingers of the hand, each perithecium contains several 4 to 8 asci.

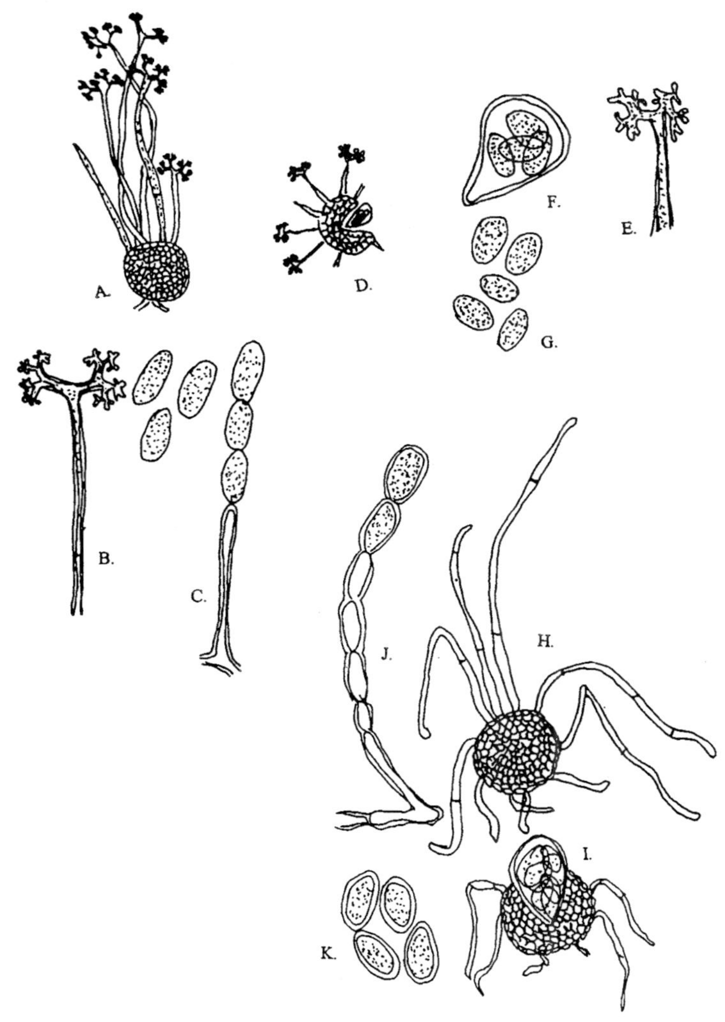


**Fig. 10.19:**

A. Conidia on conidiophore. B. Cleistothecia. C. Ruptured cleistothecia. D. Clesistothecia of *Microsphaera penicillata*. E. Dichotomously branched appendage. F. Ascus with ascospore. G. Ascospore. H. Cleistothecia of *Sphaerotheca macularis*. I. Ruptured Cleistothecia. J.Conidia on conidiophore. K. Conidia.

***Uncinula spiralis:*** This disease, was imported from America to Europe.

The fungus attacks the leaves, shoots and fruits of grape vines, both under glass and in the open air. The perithecia which contain several asci, have appendages with spiral tips.

**Pyrenomycetes**

Forms belonging to this group of fungi produce an apically ostiolate, flask shaped structure, called the perithecium, which is either embedded in the mycelium or in a stroma and is a small hard walled structure. Asexual reproduction occurs by chlamydospores or more generally by conidia which are formed either on free and separate conidiophores or conidiophores forming an acervulus or on conidiophores enclosed with in Pycnidium. The typical ascocarp develops in the ususal way with the formation of ascogenous hyphae asci with 8 ascospores and the paraphyses interspersed between them. The asci are long cydindrical and are arranged parallel on the floor and the basal sides of the perithecium. Sometimes sterile hyphal branches line the neck region of the perithecium which emerge out of the ostiole and may even close it the representative or order clavicipitales and it is important both economically and medicinally has been described here.

**Order** — Clavicipitales

**Family** — Clavicipitiaceae

**Genus** — *Claviceps.* (ergot of rye).

**Occurrence:** It occurs as a parasite on the members of Gramineae family mainly on the grasses and cereals. It attacks and fills the ovary of the host by an elongate sclerotium which completely replaces the seeds or grains. Such affected grains are called ergot grains. *Claviceps* is represented by about 12 species of which *C. purpurea* is cosmopolitan and quite commonly occuring on nearly 200 species of hosts, but its chief host is Rye and sometimes also wheat, oats, barley etc. The sclerotia which in shape, resemble a rye grain are often, linger than them. Upto 20-50% of the rye heads may shows ergot formation in a heavily infected field. Healthy as well as the ergot grains occur mixed in one and the same spikelet. During the harvesting and threasing operation in the field, the sclerotia fall on the ground where they perennate until next season and if they have reached the storage godown along with the healthy grains, they return to the field at the time of next sowing.

**Vegetative structure:** Ascospores, which are wind disseminated, they germinate on the stigma of the host flower; their germ tubes enter into the ovary and develop into a soft cottony white mycelial mat of septate hyphae. Instead of the normal grain developing with in the ovary, the entire cavity except a small portion at the upper end, is filled with these hyphae. The superficial weft of mycelium also invests the infected surface consist of elongated cells. The ovary of the infected flower is changed into a hard and horny body. This elongated mycelium is the ergot. It is composed of a compact mass of tough hyphae that in a dormant condition, survive during unfavourable periods (Fig. 10.20).

**Asexual Reproduction:** Hyphal mass which becomes convoluted in which a large number of short conidiophores form a sort of palisade layer all over the surface. These cut off ellipsoidal to ovate, hyaline and uninucleate conidia by abstriction at their apices. The conidia are formed in great numbers and collect in sweet 'honey dew' droplets at the edge of the glumes. The honey dew is secreted by the hyphal mass or by the floral nectaries due to stimulation caused by infection. This is termed as the sphacelia stage of the fungus because this manner and structure of conidia formation corresponds and was once considered to belong to Sphacelia stage on imperfect fungus. The conidia in temperate conditions are known to have long viability of several months (Fig. 10.21).

**The Sclerotium:** After the active stage of conidial formation the hyphal mass with in the ovary gets furrowed and hardens to form the sclerotium which may become apparently visible with in 1 or 2 week time after the first appearance of the honey dew. At the top of sclerotium is usually present the dried hyphal cap of the

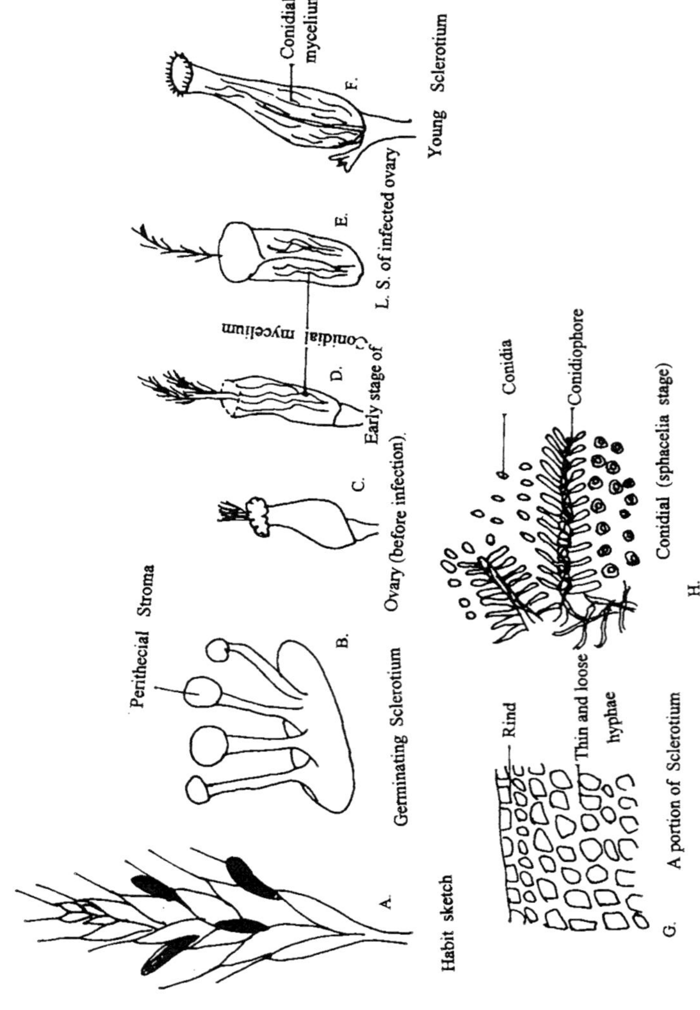


**Fig. 10.20:** Sclerotia development in *Claviceps purpurea*

Sphacelia stage in which may also be entangled the dried stamens and stigma of the flower.

The Sclerotia when young, are yellowish brown and cartilaginous but become greyish violet and horny or drying at maturity. These may measure 1-3 cm in length. A sclerotium is surrounded by a dark rind composed of thick walled compact hyphae enclosing a thin walled hyphal cortex internally. The sclerotia have a dormant period of 3 months. On germination in the presence of adequate moisture in the field, small cortical bulgings burst through the rind each of which develop into a clavate Stroma. They may number a few or more than 20 for each sclerotium.

**Sexual Reproduction** (*Stroma*): Each storma when fully developed, attains a height of 3/8 inches. It have a long slender stalk terminated by a globular head, 1-1.5 mm in diameter the stromata are yellowish white to start with, but become grayish violet at maturity with dark purple heads.

As the stromata mature, small cavities develop just underneath the surface of their heads. The hyphae form a pseudoparenchymatous sheath round these cavities and form a hyphal branch at their floor develops on ovate to cylindrical multinucleate ascogonium which is flanked on its lateral sides by one, two or more elongate or clavate antheridia, multinucleate also arise from the same branch (Fig. 10.21). The tips of the both sex organs dissolve to allow transference of the male nuclei from antheridium into the body of ascogonium. Plasmogamy occurs and the ascogenous hyphae develop upwards from the fertilized ascogonium into the cavities. A large number of elongated cylindrical asci are formed standing in a parallel manner at the floor of the cavity which corresponding enlarges into a flask shaped perithecium opening out of the stromatic head by an ostiole. These give the characteristic wart like projections all over the surface of the mature head. The sterile hyphae developing round the sex organs surround and form the wall of pertithecium. Numerous elongate paraphyses are interspersed in between the asci.

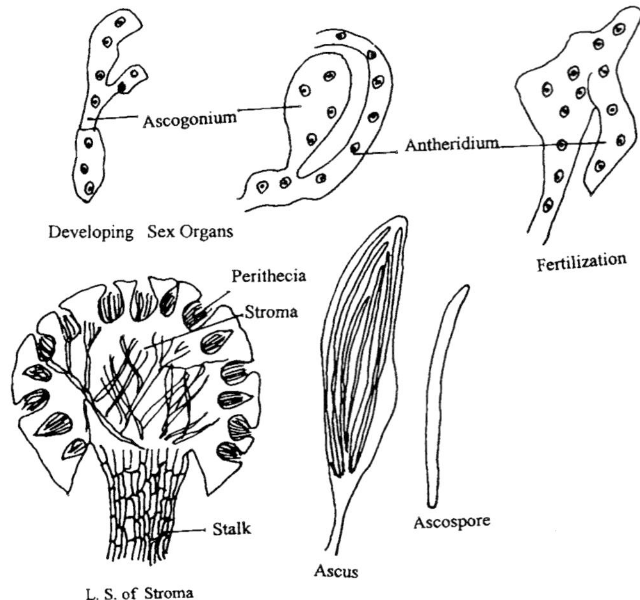


**Fig. 10.21:** Sexual reproduction in *Claviceps purpurea*

Each ascus contains 8 ascospores which are filamentous and filiform. The mature ascospores escape out violently and may be carried away by air currents to new flowers of the host. The insects may also act as their carriers if they are produced in a slimy drop over the perithecia. If the glumes are open, the ascospores germinate on the stigma or in the nectar secreted by the flower. They usually do not germinate if the investing glumes of the flowers are not open. The germ tubes of the spores fill it with a soft cottony white mycelial mat inside while, the conidia are produced externally on the surface of mycelium.

**Uses:** Sclerotia have alkaloids like ergometrive, ergotoxina, ergotamine etc. whose presence make *Claviceps* medicinally important. They yield a powerful drug which is helpful in controlling the haemorrhage during child birth and is also used as an abortifacient *i.e.* to cause rapid contraction of the uterus, thus hastening the child birth.

***Neurospora***

*Neurospora* is widely used in genetical and biochemical studies, particularly in genetical and biochemical studies, particularly eight spored heterothallic species namely *N. crassa* and *N. sitophila* and four spored *N. tetrasperma*. The species of *Neurospora* have been proved to be useful in biochemical and genetic research due to their following attributes:

(i) Wild strains are easily isolated from burnt ground, charred vegetation, warm humid environments and bakeries.

(ii) Mutation can be induced readily by irridiation of conidia.

(iii) Growth and sexual reproduction is rapid.

(iv) Wild strains have simple nutritional requirement.

(v) Tetrad analysis by ascus dissection as relatively easy.

(vi) Uninucleate cells, with in 24 hours mycelium developed branched chains of numeous multi-nucleate (oidia) which are readily dispersed by wind.

(vii) In *Neurospora* carotenogenesis is an aerobic process in absence of light, hypothetical stable photo produced—X that is oxidised and leads to subsequent morphological effects.

*Neurospora* is commonly known as bread mold or red bread mold, because it can easily infests bakeries and causes severe damage. Mycelium is made up of branched hyphae. Aerial hyphae form a mass of mycelium and orangish pick colour conidia are present on it. Conidia are present on the conidiophores (Fig. 10.22). Conidial stage of Neurospora is known as *Monilia sitophila*. Conidia are mutlinucleate but with this uninucleate conidia are also present. Sexual reproduction takes place by means of Ascogonium which is represented by protoperithecia, in each of which a multinucleate ascogonium is embedded. The ascogonia produce long hyphae branches that function as Trichogynes, antheridia are not produced. Male elements are represented by microconidia produced in chains on microconidiphores (Fig. 10.22B), a conidium or a germ tube, can also supply nuclei to the receptive trichogynes. When perithecia (Fig. 10.22C) is matue it is dark coloured, pyriform and beaked and contain octosporous asci, but paraphyses are absent. Spores are dark brown or black with nerve like ribs on the outer wall that characterize the genus *Neurospora*. In each Ascus eight ascospores (Fig. 10.22D) are present among them four are of one mating type, other four are poduce a self fertile mycelium. *Neurospora* belongs the family-Sordariaceae.

*Order* —Sordariales.

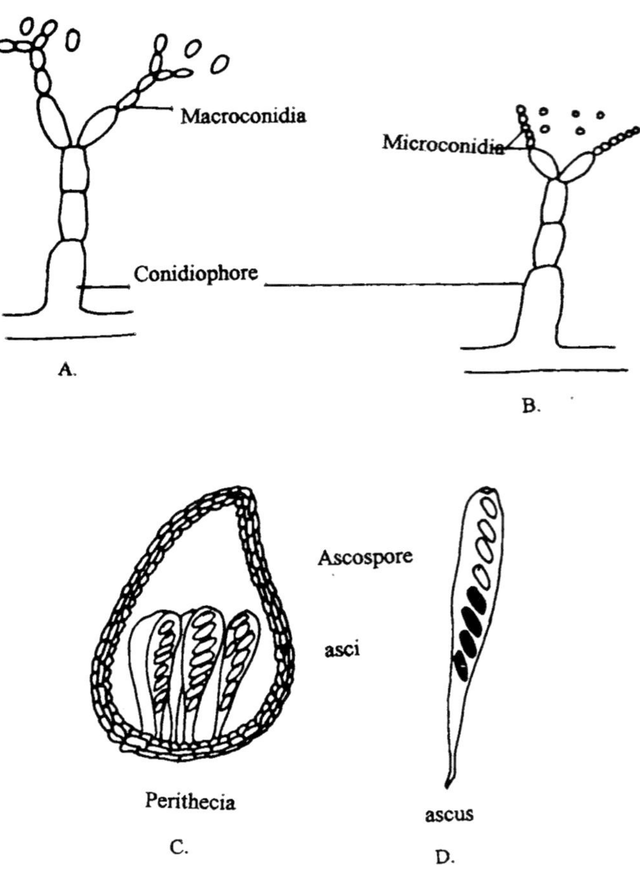


**Fig. 10.22:** Conidia and Perithecia of *Neurospora*

### *Xylaria*

It belongs to family Xylariaceae and order xylariales. It is a saprophytic fungi, which is generally found in lawn, saw dust, trees, leaf mold and into the soil. Stromata of *Xylaria* are large in size, produced in clusters. Stromata produce perithecia over the entire fertile portion. Stromata are corky, leathery or woody in nature. Generally it is dark brown or black in colour from the outside, while white fom the inne side. In the *X. polymorpha* stromata is club shaped while in *X. hypoxylon* it is tall, slender, sub cylindrical or flattened.

### Discomycetes

Forms belonging to this group of fungi produce a disc shaped or cup like ascocarp called the apotheium. Some of these fungi are the most easily recognized of all ascomycetes because of their large and colourful ascocarps, which often take on the forms of cups, saucers and cushions. Apothecium exhibit various bright colours *i.e.* orange, red, yellow and sometimes even black.

In habit these forms are saprophytic growing on organic matter containing abundant moisture; the common substrata being rotten wood logs fallen and decaying leaves and fruits, animal dung heaps and also on the ground.

The Discomycetes have been subdivided into 3 orders, Pezizales, Helvellales and Tuberales Discomycetes have been divided into two groups (*i*) operculate (*ii*) inoperculate.

(*i*) Operculate discomycetes have completely closed, hypogeous ascocarps. In these ascocarp, asci line internal chambers of ascocarp tissue and the ascospores have passive spore discharge with dispersal.

(*ii*) In inoperculate asci do not open by a definite pore but they burst irregularly at the tip to release the spores.

**Pezizales:** Members of this order have large cup or saucer shaped apothecium, which may be a closed structure. They occur mostly as saprophytes on soil, dead and decaying wood or humus and dung piles. The mycelium is well developed forming a complex hyphal system in the substratum which acts as the absorptive organ for the aerial apothecium which is generally unstalked. Sex organs comprise the ascogonium bearing terminally a long or short trichogyne coiling round an elongate, cylindrical antheridium. The asci contain 8 ascospores which are set free by their operculate dehiscence. Most species exhibit degeneration in the sex organs especially in regard to the structure and function of the antheridium. Asexual spores like conidia, chlamydospores and oidia are usually formed. Pezizales can be grown in pure culture on defined media containing a sugar and mineral salts. Ascospores placed on the medium germinate to produce a rapidly spreading and branching mycelium of large hyphae.

***Peziza:*** This genus includes many 'cup fungi' one of the species *P. coccinea*, is conspicuous in woods in early spring, attacked by its short stalk to rotting sticks, and its upper part forming a cup about an inch in diameter, the whole looks like a small scarlet wine glass. *Peziza aurantiaca* with smaller cup and scarcely any stem occurs in similar situations in autumn. *P. stercorea* an orange coloured form is common on cow dung, and a brown species is frequently seen on manure heaps and rotting leaves. They are parasitic also in nature, prey on the living tissue of larch, beach, oak and other trees.

**Mycelium:** Vegetative part of the fungus is white and filamentous, consisting of branched, septate hyphae, it ramifies in the organic matter on which it lives, this is the case of parasites being a living host. Stroma is the modified portion of the mycelium concerned with the reproductive functions.

**Asexual Reproduction:** This function is comparatively rare in *Peziza*. It has been observed in some species in the lab by keeping plants killed by the fungs in a moist chamber. Creeping hyphae develop and give off lateral branches with swollen ends covered with elliptical colourless conidia. There is no record of the production of conidia out doors, but quite frequent (Fig. 10.23).

**Sexual Reproduction:** Sexual organs, antheridium and ascogonium are the apical cells of thick hyphal branches, which

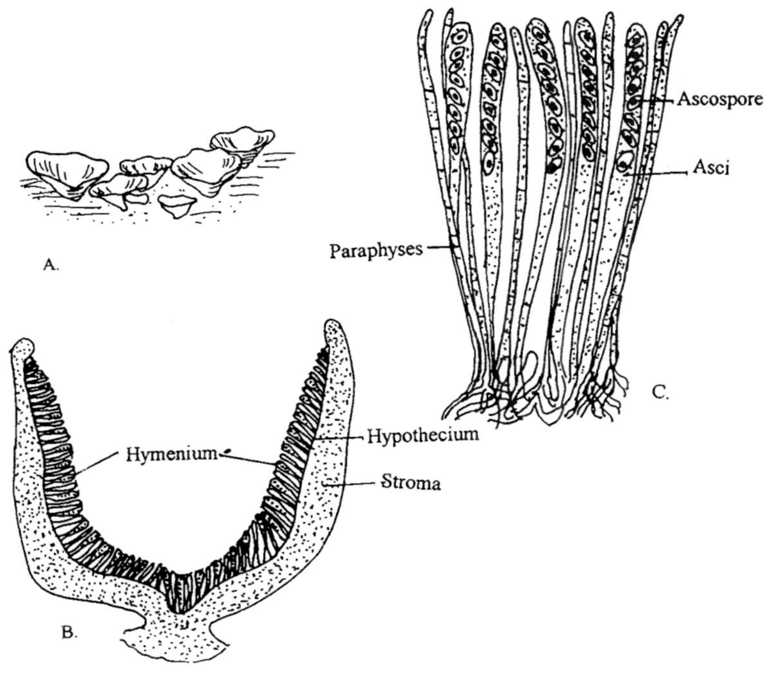


**Fig. 10.23:**

A. Cup of *Peziza*. B. Apothecium of *Peziza*. C. Detail structure of Asci part.

arise vertically from the substratum, and stand side by side. At the apex of the ascogonium there is a receptive out growth the trichogyne, by way of which the male gamete passes over the female gamete and effects fertilization. There after the ascogonium produces many ascogenous branches, the ends of which are cut off to form asci. This sexual process takes place on that part of the mycelium which forms the fruit body or cup and the ultimate position of the asci in this fruit body is a characteristic feature (Fig. 10.24).

A fruit body of this particular form and structure is termed as apothecium. When quite young it is globose and closed, but it

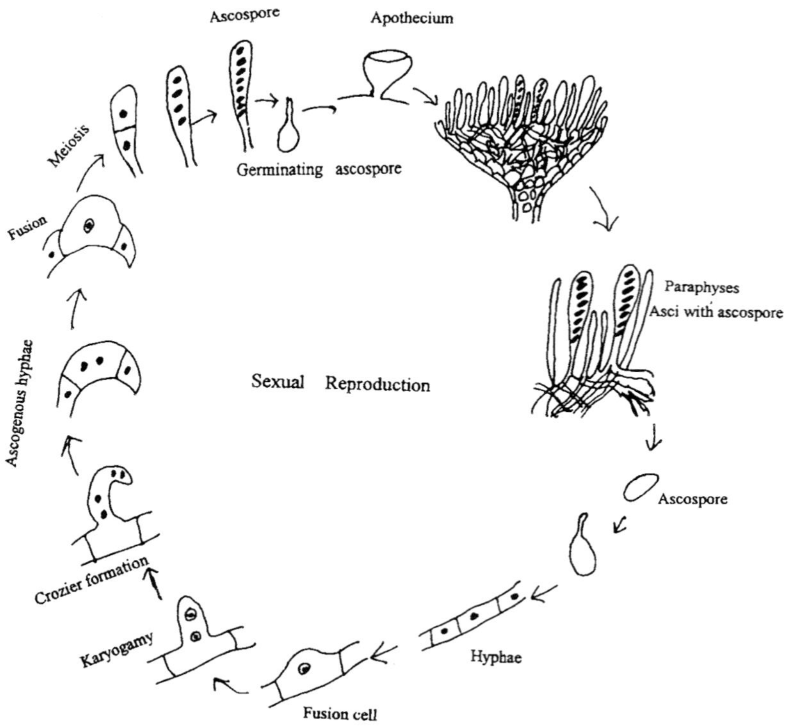


**Fig. 10.24:** Sexual reproduction in *Peziza vesiculosa*

expands as it matures until it is cup or saucer shaped. At this stage apothecium shows the following features:–

1. The smooth upper surface, or ascigenal layer.
2. The lower surface, wall or peridium, in the sessile forms this bears fine hyphae filaments which connect with those in the substratum, in the stalked forms it narrows into the stalk or stipe.
3. The interwoven hyphae between (a) and (b), forming a pseudo-parenchymatous tissue and merging into the peridium imperceptibly.

The ascigenal, a spore bearing layer, which forms the lining of the cup, consists of a single layer of closely packed, club shaped asci, produced by the branches of the fertilized ascogonia,, with paraphyses between them. Each ascus contains eight oval ascospores. Apothecium corresponds to a single ascocarp or perithecium split and opened out, so that the spore bearing layer is exposed. When the apothecium is expanded the mature asci absorb moisture and swell, around the edge of the ascigenal layer is the slightly developed margin of the apothecicum, creates a pressure on the asci when these expands, and aids in the forcible expansion of the ascospores.

The free ascospore, under suitable condition germinates, emitting a germ tube, which develop into a mycelium again.

### *Ascobolus*

**Habit and occurrence:** *Ascobolus* is a coprophilous saprophyte occurring on animal dungs such as cow, horse and sheep. It may also sometimes grow on soil rich in humus. *Ascobolus carbonarius* occurs on burnt ground among charcoal. The apothecia may become one inch or more in diameter, are generally purpulish brown when mature and become greatly elongated and laterally distended due to absorption of copious water from the substratum. The hymenium of the mature fructifications, if tapped releases a dust of ascospores. *Ascobolus immersus* have a very large size ascospores measuring upto 75μ to 35μ.

**Vegetative structure:** The mycelium ramifies in the substratum forming a complex hyphal system which acts as the absorptive organ for the aerial apothecium, developing a sessile discoid structure directly over the substratum. The hyphae are branched and septate, the cells being multinucleate. The hyphal tangle in the general body of the apothecium forms a pseudo parenchymatous tissue.

In some species, sometimes oidia and conidia are produced on small branches of the hyphae. Chlamydospores may also be formed in some cases. These structure may develop either singly or in chains.

**Sexual Reproduction:** The sex organs or ascogonium and antheridium which appear generally 4–6 days after the germination of the ascospore. These exhibit great variations in their form and behaviour in the different species of *Ascobolus* so far as the copulation between the female and male branches and degeneration in the structure of the latter is concerned.

***Ascobolus scatigenus:*** A common form growing on the moist dung piles of the cows and horses, exhibits a typical case of sexual reproduction, where the two sexual branches arise vertically up from the hyphal base. It is a monoecious, self sterile and heterothallic species. Antheridium as a cylindrical or clavate branch or younger hyphae and remains vertically erect. The female branch develops on older hyphae, elongates and terminally cuts off a long trichogyne which by septation, becomes a several or more celled structure. Simultaneously the cell below the trichogyne enlarges and swells into a globose ascogonium. Both the sexual branches are multinucleate and are supported on a multicellular stipe. Searching for the antheridium, the trichogyne reaches the antheridial tip and coils round the body. The separating wall between the two dissolves and the male nuclei pass into the apical cell of the trichogyne, from where they reach the ascogonium, the septa between the trichogyne cells being perforated. With the entry of the male nuclei in the ascogonium, the female nuclei divide and becomes arranged on the periphery. Where the two nuclei, one from each form pairs. Soon the ascogenous hyphae are budded out from the fertilized ascogonium which are narrow and elongate to start with. Later their branches fork out and form the terminal cells, carrying the dikaryon, the asci containing 8 ascospores are formed in the usual way. The ascospores are one celled, double walled ellipsoidal or spherical, purplish or dark brown and escape out through the apical lid. These ascopores germinate by a germ tube to form the new mycelium and apothecia.

### *Pyronema*

*Pyronema* is a saprophyte occurring gregariously on ground which has been burnt by fire. It also occurs on steamed soil *i.e.* on such pots which have been autoclaved for sterilising their soils and even at the bottom of damp piles of charcoal. The apothecia are convex, lentil shaped, 1–3 mm, wide and the adjacent ones become confluent with each other. Because of the profuseness and deep rose red, pink, orange or flesh colour of these fructifications their growth becomes a conspicuous feature for a few days.

**Vegetative structure:** The mycelium is superficial, colourless or whitish and consists of thin and septate hyphae whose cells are multinucleate. This fungs does not reproduce asexually.

**Sexual Reproduction:** Sexual reproduction is typical of ascomycetes taking place by a very simple ascogonium and an antheridium. From the underlying mat of the mycelium. Some hyphae arise upwards and branch repeatedly so as to form a tuff or rosette of hyphal branches. Sex organs are developed terminally over these. The female branch is the first to develop and arises as cylindrical or oval single celled ascogonium which bears, at its apex, a curved unicellular trichogyne; both these are multinucleate, having 100–200 nuclei. Antheridium develops as a clavate or as an obvate structure either from one of the cells of the multicellular stipe of the ascogonium or from a separate branch of hypha. The antheridium too, is multinucleate containing 100 or more nuclei. As the sex organs are maturing, the apex of the trichogyne connects the antheridium and may coil or twist around it. The separating wall between the two dissolves. Meanwhile all the nuclei in the trichogyne and some of the nuclei both in the antheridium and the ascogonium degenrate the majority of the nuclei migrate through the trichogyne into the ascogonium where in the female nuclei have collected around a central hollow sphere. Nuclear pairing takes place between the male and female nuclei.

After some time about 10–20 buds develop over the fertilised ascogonium; these elongate to form the ascogenous hyphae separation occurs and the basal cell of these hyphae have several nuclei while terminal cells are dikaryotic. They branch profusely and the dikaryons increase by conjugate division. After the crozier formation and fusion of the two nuclei in the young ascus. 8 spored asci are produced in the usual way which are intermingled with long and septate paraphyses. Basal cell which support the group of male and female organs grow out and branch upwards to form external hyphae as looser layer of sheath surrounding the group of sex organs. Hyphal tangle of these sex organs become inter winded and form several apothecia they become confluent with each other signifying the name *Pyronema confluens*.

The ascospores on liberation, germinate to form the new mycelium.

**Development of apothecium:** There are two methods of development of the apothecium and both occur in the genus *Ascobolus*:–

(1) Angiocarpic

(2) Gymnocarpic

**Angiocarpic:** Apotehcium, when young is closed. Sterile hyphal branches, that arise from the stipe cells of the archicarp, surround the fertilised ascogonium from which the ascogenous hyphae develop vertically upwards. The outer cortex of this apothecium initialy is thick walled. While the inner hyphae are thin. Towards the upper portion, a mucilage cavity develops by the dissolution of hyphae and the palisade layer comprising the asci and paraphyses originates at the base at this cavity. This layer is the future hymenium which expands and ruptures the covering cortex over it. The disc or the cup enlarges laterally and the asci starts maturing centrifugally.

**Gymnocarpic:** The apothecium remain open from the very beginning. The ascogenous hyphae remain separate from the sterile hyphal sheath and the hymenium comprising the asci and paraphyses remain exposed at the upper surface when the apothecium is young.

**Structure of Apothecium:** Asci when mature, along with paraphyses project beyond the general level of fertile layer--hymenium of the apothecium. The ascus is an elongate, clavate sac like structures enclosing 8 ascospores which are uniseriate or tending to become biseriate. Each ascospore is usually ovate to elliptical, sometimes spherical double walled, uninucleate and purplish to brown, when mature. The paraphyses one slender, narrow, sterile hyphal propagations from the hypotheium or subhymenium which is a layer of compactly interwoven hyphae underlying the hymenium. The paraphyses may be septate and multinucleate. The general body of the cup is composed of interwoven mesh of hyphae forming a pseudoparenchymatous tissue called the trama. The hypha towards the exterior have thick walled cells forming a distinct peridium while those at the base form an absorptive rhizoidal system going with in the substratum.

The ascospores when mature, escape through a lid at the apex of the ascus and germinate into a new mycelium by putting forth a germtube under favourable conditions.

**Family Morchellaceae:** Members of Morchellaceae are characterized by large often stalked apothecia, mostly with a sponge like or bell shaped piles. It includes the morels and the bell morels.

**Genus *Morchella*:** *Morchella* is commonly known as the morel fungus or sponge-mushroom because of its pileus being conspicuously ridged like a sponge. It is a saprophyte occurring in deciduous forests. Morels are whitish grey to dark brown in colour depending upon the species and their age. Their height ranges from one inch to 4–5 inches. The fertile head or the pileus is supported over a thick stipe and is characteristically pitted. It is coarsely thrown into irregular longitudinal and transverse folds to form conspicuous ridges and enclosing deep regions or cavities which are lived by the hymenium, its fertile surface is greatly increased. *M. esculenta* is one of the species with large apothecia. It is supposed to be best in North America.

**Vegetative structure:** The mycelium ramifies in the ground comprising multicellular hyphae. It does not generally reproduce by asexual means.

**Sexual Reproduction:** Typical sex organs are not formed and the apothecia are formed as a result of somatogamy copulation and fusion between two cells of the vegetative hyphae. The copulation hyphae branches are wider and distinct from the neighbouring ordinary hyphae. Branched ascogenous hyphae grow upwards from the cells of these hyphae in which the nuclei lie in pairs and the asci are formed after the usual method of crozier formation. The hymenium layer comprises asci which are long clavate and operculate. Each ascus encloses a large, ovate, hyaline ascospores arranged uniseriately within the ascus. The ascospores at maturity may become 8 nucleate. Several elongate paraphyes are interspersed between the asci. The ripe ascospores are discharged between the asci. The ripe ascospores are discharged explosively with such violence that they are ejected several centimeters away. Thus in the hymenium layers that line the deep lacunae of the pileus, the discharged ascospores may be shooting against each other. The ascospores germinate by putting forth germ tubes to form a new mycelium.

*Morchella* is edible and is a delicacy. It is cultivated commercially also. The fresh or dried ascocarp, if mixed in the suitable soil will from new crops next season. In India, morels are cultivated commercially in the Kashmir valley.

# 11. BASIDIOMYCETES

Fungi belonging to the division Basidiomycota is known as **Basidiomycetes**. They are large and diverse lot and include forms commonly known as mushrooms, boletus, puffballs, earth stars, stink horns, birds nest fungi, Jelly fungi, bracket or shelf fungi, rust and smut fungi.

Basidiomycetes are characterized by exogenous production of spores, termed **Basidiospores** on specialized structure called Basidia. These are sexually produced spores just like asci and ascospores of ascomycetes. A young basidium is binucleate where the karyogamy takes place quickly followed by meiosis and resulting in the development of 4 spores that are exogenously formed on short **sterigmata**. The number of basidiospores may however vary and in some species only 2 spores may be produced while in others particularly in the order **ustilaginales** as a whole, the number of basidiospores, formed on a single basidium is indefinite. The Basidiomycetes have character of higher fungi i.e. presence of septate mycelium and complete absence of motile phase in the life cycle. Plasmogamy occurs early in the life history but karyogamy is delayed until the development of the basidium.

**Habit and occurrence:** Basidiomycetes are mainly terrestrial in nature, occurring as parasites or saprophytes. Among the parasitic forms, there are two main groups of fungi that are commonly known as rust and smuts. These are responsible for causing serious damage to our food crops. The higher basidiomycetes, sometimes also designated as the true basidiomycetes. They are mainly saprophytic in nature and are represented by forms like the mushrooms, toad stools, bracket fungi, puff balls, Jelly fungi etc. Some species of these are parasitic fungi causing diseases of forest trees.

**Vegetative Structure:** Mycelium consists of well developed branched and septate hyphae that grow or in the substratum but can not draw nourishment from it. In case of the parasitic forms the hyphae within the host tissue may either be intercellular sending haustoria in adjacent cells or intracellular. The cells of the hyphae may be uninucleate and binucleate, Recent studies on the structure of septa in higher fungi have revealed that in the basidiomycetes, the septa are generally broad in the middle surrounded by a double membrane **parenthesome** on each side and forming a more or less barrel shaped structure with a minute pore in the centre open at both ends (Fig. 11.1). This is known as the **dolipore septum**, through which the cytoplasmic continuity can be maintained between adjacent cells but there is no migration of the nuclei, while in ascomycetes, the septum tapers centrally into a simple clear channel from which all material including nuclei can freely pass between cell to cell.

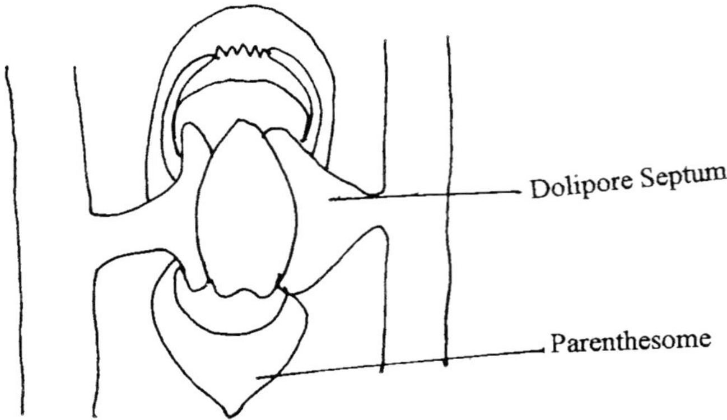


**Fig. 11.1:** Dolipore septum in Basidiomycetes

From the germination of the basidiospore until the end of the vegetative phase and production of the basidia and basidiospores. The mycelium passes through two or sometimes three distinct phases: These are designated as (1) Primary (2) Secondary (3) and Tertiary mycelium. The primary mycelium is known as haploid 'haplophase or the monokaryon mycelium and the secondary as 'diploid' diplophase or the dikaryon (Fig. 11.2).

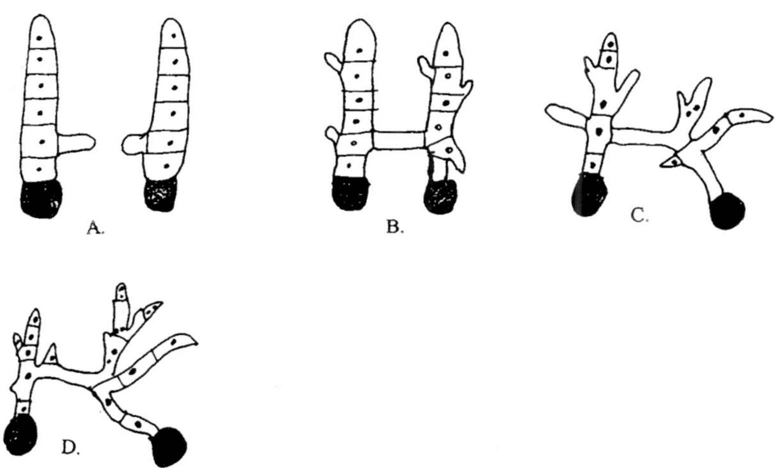


**Fig. 11.2:** Stages in the formation of secondary m.ycelium through somatogamous copulation between two unınucleate cells of primary mycelium of opposite strains. (A-D)

The primary mycelium develops from the germination of the basidiospores. It consists of uninucleate cells and is short lived. It may be multinucleate in very early stages when due to the repeated and quick division of the basidiospore nucleus, large number of nuclei may be seen in the germ tube; but this multinucleate phase is very short lived and soon the mycelium gets divided into uninucleate cells by septation. In others, the uninucleate cells are formed from the very beginning. The primary mycelium usually has a limited growth and does not normally give rise to the sporophore, basidia or basidiospores. Sporophore is produced rarely, but basidia formed are only few and usually sterile. They may sometimes develop 2 uninucleate basidiospores both of the same sexual strain. The monokaryon mycelium frequently produced oidia which are small uninucleate spores formed apically on branched or unbranched oidiophores. They are cut of serially from apex towards the base leaving behind short stub of the oidiophores. The oidia sometimes are held in a drop of liquid. In the rust, special uninucleate male cells are termed as spermatia are produced in flask shaped ostiolate structures called spermogonia. These also develop from the monokaryon mycelium.

The secondary mycelium develops from the primary mycelium and consists of binucleate cells. The process of development of binucleate condition from the uninucleate condition of the primary mycelium is termed diploidisation or diakaryotisation and the binucleate mycelium is known as dikaryon. Many species are heterothallic and the secondary mycelium is initiated only when a compatible hypha or spore of a different sexual strain comes in contact with the primary mycelium of an opposite strain.

Binucleate condition may originate by one of the following ways.

1. By conjugation either directly or through copulation tubes of two basidiospores, when the nucleus from one migrates into the other, thus imitating the binucleate condition at a very early stage in the life history. With repeated conjugate division of the nuclei, the mycelium that develops is dikaryotic through out.
2. By fusion between the oidia or spermatia of one primary mycelium with another primary mycelium of opposite sex. The wall between the oidium or the spermatium and the hyphal cell dissolved and the nucleus from the former migrates into the latter. Once the binucleate cell has been initiated it may grow out into a dikaryon hypha by elongation and repeated conjugate divisions of the two nuclei or the introduced nucleus may divide and one of the daughter nucleus passes into the adjacent cell through he shifting of the septum. This process continues until the whole hypha is dikaryotised (Fig. 11.2).
3. By union of two sexually opposite monokaryon hyphae, they come in contact with each other at one or more places the wall between the cells dissolve and the nucleus from one cell migrates into the other. Cells of both the mating

hyphae may be diploidised simultaneously from the nuclei of other strain. Once the binucleate cell has been initiated, the rest of the development of the dikaryon occurs by either of the methods discribed in the previous section.

4. By union of an oidial mycelium of one sexual strain with the primary mycelium of another strain, or by union between the two oidial mycelia of opposite sexes or also by fusion between an oidium and a cell of the oidial mycelium both belonging to different strains.

In higher Basidiomycetes a special process is associated with the diploidisation by which the nuclei arising from conjugate division of binucleate cell are separated in the daughter cells. This takes place through the formation of special structures known as clamp connections. Presence of clamp connections is an indication of its being a secondary mycelium but its absence does not indicate the monokaryon nature. These may be present at every septum or may be absent from some part of the hypha. They occur more regularly in slender hyphae being absent from parts where the cells are broader. The details are as follows.

The two nuclei of a the dikaryon cell lie in a short distance apart in the longitudinal axis. A short lateral pocket like outgrowth develops between them medianly and forms a downwardly projecting hook like structure the clamp or the pouch (Fig. 11.3A-E). The nuclei now come closer and lie more or less parallel to each other near this out growth. Both the nuclei divide simultaneously, undergo conjugate division in such a manner that the spindle of one nucleus lies obliquely in the lateral pocket and the spindle of the other nucleus lies parallel to the longitudinal axis of the cell. Septa are laid down separating the daughter nuclei. Out of the obliquely oriented spindle, one nucleus remains in the pouch cell and the other in the hyphal cell. Similarly one of the daughter nuclei of the longitudinally oriented spindle remains in the upper cell. While other goes into the lower cell. At this stage the upper cell has two nuclei and the cell below it and the lateral pouch cell both have one nucleus each. The lateral clamp curves further so that its tips comes in contact with the hyhal wall of the lower cell. The

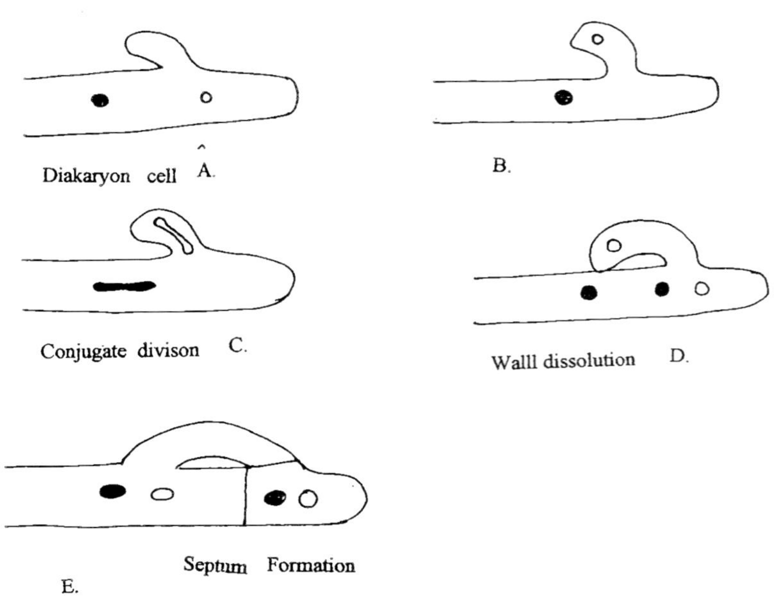


**Fig. 11.3:** Stages in the clamp connection

wall between the two dissolves and the nucleus from the pouch migrates into the cell thus making the lower cell also binucleate— One of each sex. This entire process may be completed in 23–45 minutes. The lateral pocket which has served as a by pass or a bridge through which one of the daughter nucleus has been transferred from one cell to another is known as **clamp connection**.

It is the binucleate secondary mycelium that generally bears the basidia and basidiospores. However in many higher Basidiomycetes there occur well developed and conspicuous sporophores where the **secondary mycelium** forms organised and differentiated complex tissue. This phase of the development of sec mycelium in the sporophores is sometimes designated as the **tertiary mycelium**. The cells of the tertiary mycelium are also binucleate.

**Asexual Reproduction:** Asexual Reproduction in basidiomycetes takes place by fragmentation of the hyphae, budding, conidia, oidia, chlamydospores or gemmae and formation of arthrospores.

Conidia formation is of common occurrence in the smuts and they may either be uninucleate or binucleate depending whether they are formed on the monokaryon or the dikaryon mycelium. They arise on short generally unbranched conidiophores conidia are also known to be budded off from the basidiospore. Among the rusts, the conidia that are of common occurrence on the dikaryon are termed as **uredeniospores** which are binucleate structure. Some of the higher basidiomycetes also produce conidia (Fig. 11.4).

Oidia are of common occurrence in many of the higher basidiomycetes. These are produced on special branches called **oidiophores**. Oidia are cut off in succession from the apex of the oidiophores and thus a single oidiophore produces a large number of oidia. These germinate to give rise to a new mycelium. They are also known to play an important role in the dikaryotisation of a primary monokaryon mycelium (Fig. 11.4).

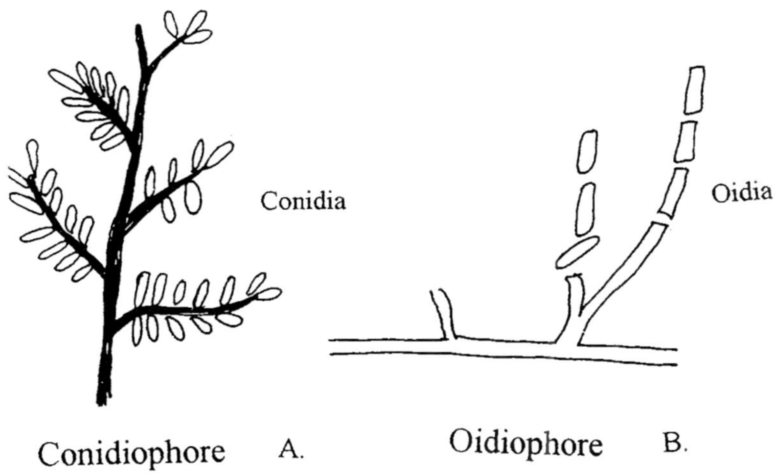


**Fig. 11.4:** Asexual reproduction in Basidiomycetes A. Conidia. B. Oidia.

**Arthrospores** are unicellular mycelial fragments that develops in some basidiomycetes. The hyphae break up into single celled pieces which develop a thick wall without rounding up. The arthrospores may be uninucleate or binucleate depending on whether they are produced from the primary or secondary mycelium. They germinate by a germ tube and give rise to a new mycelium.

**Sexual Reproduction:** In the Basidiomycetes, there is generally a complete absence of functional sex organs except in the rusts, where some specialised structure functioning as sex organs like the receptive hyphae (female) and spermatic (male) are found. Generally, these occurs typical somatogamy in which the sexual reproduction takes place by fusion between two somatic hyphae or between an oidium and a cell of the somatic hyphae both belonging to opposite sexual strains. Thus the somatic structure have taken over the function of sex organs. Plasmogamy occurs early in the life history and the dikaryophase represented by the binucleate secondary mycelium forms the dominant phase.

Sexual reproduction in the Basidiomycetes is generally initiated by any of the following methods which may occur by fusion between.

1. cells of two primary mycelia
2. an oidium and a cell of primary mycelium
3. an oidium and a cell of the oidial mycelium
4. cells of two oidial mycelia.

In each of these cases the mating structure belong to two opposite sexual strains. On coming in contact with one another the seprating wall between the two dissolves and the plasmogamy results in diploidisation in which the cell becomes binucleate. Once the binucleate condition has been achieved by any of the methods described earlier the rest of the secondary mcelium develops and nearly all the cells of the mycelium become dikaryotic. The secondary mycelium initiates the formation of sporophores in the higher basidiomycetes. The dikaryophase continues its growth for major part of the life history of the fungus and ends with the development of basidia and basidiospores. The dikaryon initiated earlier as a result of sexual reproduction, finally culminates in the fusion–Karyogamy or the two nuclei in the young basidium followed quickly by meiosis and resulting in the formation of 4 haploid nuclei after two divisions. Consequently, 4 basidiospores develop exogenously over the basidium. Thus in the basidiomycetes also, inspite of the lack of functional and traditional sex organs the three essentials of a typical sexual reproduction i.e. plasmogamy. karyogamy and meiosis takes place in a definite sequence and at definite stages in the life history. While some varieties may be achieved in different species, there is a complete consistency with regard to karyogamy and meiosis that takes place only in basidium (Fig. 11.5 A-F).

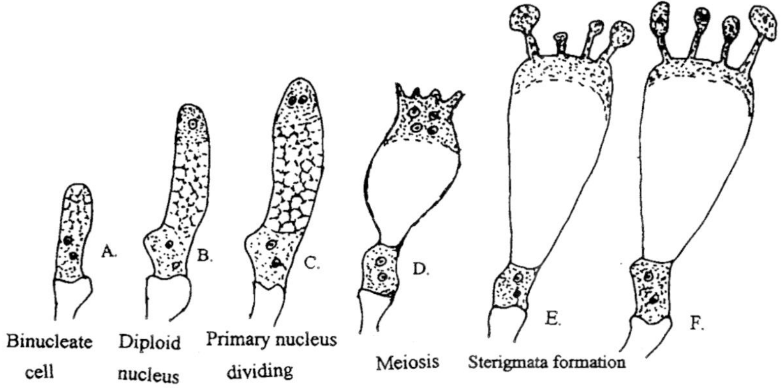


**Fig. 11.5:** Successive stages in the development of holobasidium

In some of the higher basidiomycetes, plasmogamy may not occur at all in the life history as the basidispores that develop are binucleate from the very beginning and on germination, result in the development of a binucleate mycelium by repeated conjugate division of the nuclei. In such forms, the primary mycelium may not be formed at all. However, the karyogamy and meiosis takes place in the young basidium.

**Basidium:** Binucleate secondary mycelium that initiates the development of the fruit bodies, ultimately forms the basidia. The terminal cell of the binucleate hypha that has to develop into a basidium, is cut off from the remaining part of the hypha by a septum and is usually associated with a clamp connection. This terminal dikaryon cell is young basidium which is narrow and elongated to start with but soon starts enlarging into a relatively large and clavate structure. The two nuclei fuse in the young basidium resulting in Karyogamy and the fusion nucleus immediately undergo two successive divisions resulting in the formation of 4 haploid nuclei by meiosis. The reduction occurs either in the first or in the second meiotic division. As these nuclear changes are taking place internally the basidium enlarges its apical and becomes broader and puts forth 4 small processes the sterigmatas. The tips of the sterigmatas gradually enlarge becoming globose to form the basidiospore initials. One nucleus along with some cytoplasm is squeezed into each of the basidiospore initial through the narrow opening between the sterigmata and the basidium. The spore initials enlarge and develop into a characteristic 4 uninucleate haploid basidiospores. There is a great variation observed in the form and structure of the basidium and also in the shape and size of the sterigmata in the whole group of basidiomycetes. Basidia become septate either transversely as in Ustilaginales and Uredinales or longitudinally and sometimes obliquely as in Tremellales. In such cases, after karyogamy in the young basidium the nuclear division is followed by separation which divides the basidium into 4 cells, each uninuceate lateral or terminal sterigmata are given out. Whose apex the basidiospores develop in the usual manner, either more in number or only four. The septate basidium is sometimes termed as phragmobasidium, non septate is known as 'holobasidium' (Fig. 11.6).

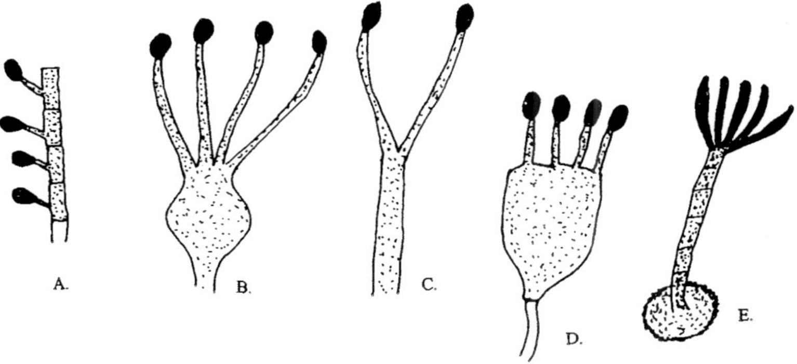


**Fig. 11.6:** Various kind of basidia

A. Stichobasidial type. B. Chiastobasidial type. C. Tuning fork type. D. Holobasidium. E. Stichobasidial type with terminal cluster of septate, sickle shaped sporidia.

In the rust and smut the nuclear fusion takes place in the young basidium occurs in the special thick walled structures called 'teliospores' or brand spores. These on germination give rise to the basidium in which meiosis occurs and later bear the basidiospores. This structure one time is also known as metabasidium or promycelium.

**Basidiospores:** Generally, 4 basidiospores are formed on each basidium but in smuts, the number of basidiospores are indefinite, they are termed as sporidia. Basidiospores are commonly borne over sterigmata and are normally unicellular with one haploid nucleus. They are very variable in shape and colour, the latter ranging from hyaline to black, while the shape may be globose, oval or elongate. Mostly they are thin walled but sometimes thick walled.

Basidiospores are attached obliquely on the sterigmata and in most of the basidiomyetes, a small liquid drop appears just before their discharge over the hilum which is a lateral projection at the point of attachment of the spore of the sterigmata. Liquid drop is formed with in the fragile extension of the sterigma-wall around the base of the basidiospore. This droplet increases in size to gradually attain ¼ of the size of spore diameter. The spore is shot off from the strigmata with considerable force upto a distance of 10–20 times of its own length and carries a droplet along with it. Thus there is explosive mechanism for the discharge of basidiospores (Fig. 11.7 A-I). In the presence of suitable moisture and temperature they germinate by a germ tube and give rise to the uninucleate primary mycelium.

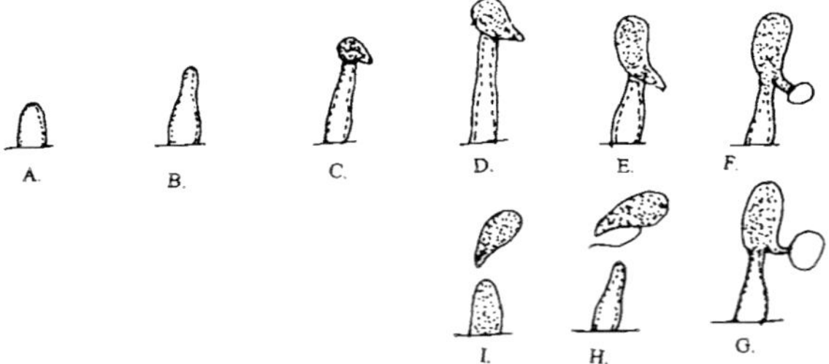


**Fig. 11.7:** Stages in the development and discharge of basidiospore

In some Basidiomycetes, basidiospores may be binucleate at maturity. This may happen in one of the two way either.

(1) the single nucleus of the basidiospore may divide into two before its discharge from the basidium. Such basidiosproes although binucleate on germination give rise to uninucleate mycelium by septal development during or other germination.

(2) after the meiotic division in the basidium into 4 nuclei only 2 basidiospores are formed into each of which two nuclei migrate. Such basidiospores on germination give rise to binucleate mycelium which completes the life cycle without mycelial contact and plasmogamy.

**Basidiocarps:** Fruit bodies of the basidiomycetes are commonly called the basidiocarps or sporophores. Except the rust and smut the rest of the basidiomycetes produce distinct and quite conspicuous basidiocarps, which are variable in shape, size, texture, colour and their longevity. The variation in size of the fructifications may range from minute and microscopic that reach a size of 3 feet or more in diameter. They may be gelatinous fleshy, papery, leathery corcky or even woody in texture.

Basidiocarps may remain open from the beginning with their fertile layer hymenium exposed or they may open only at maturity or may remain closed. In last category, fructification is usually hydrogenous, fertile inner layer is known as 'gleba' and outer sterile and relatively hard covering forming the rind of the peridium.

Classification of basidiomycetes is as follows: (Fig. 11.8).

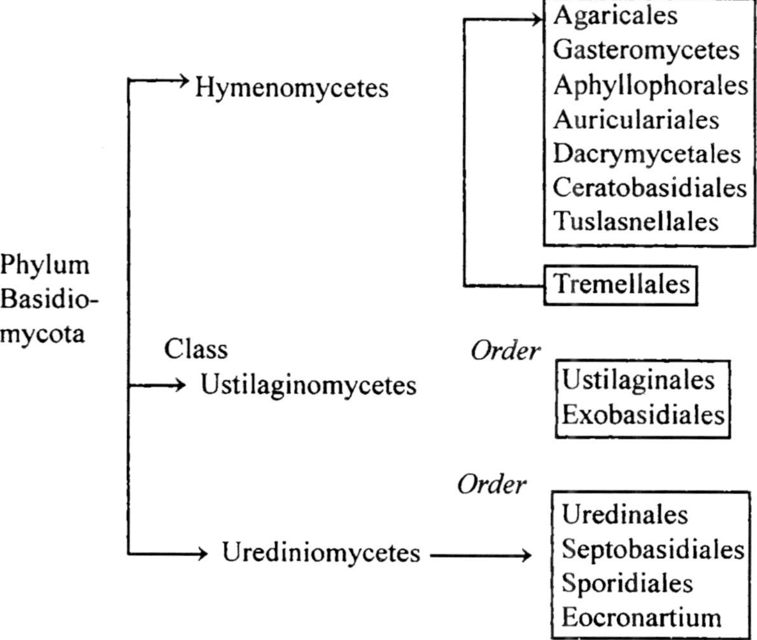


**Fig. 11.8:** Basidiomycete evolution Based on morphological characters and rDNA sequence analysis
(Source: Alexopoulous, Mims and Blackwell, 1996)

Most of the basidiocarp, whatever be their form and structure are lived in a characteristic manner in some part of their body by the fertile layer–The hymenium. Hymenium carrying the basiodospores but in addition a number of sterile hyphae termed as paraphyes also occur intermingled with the basidia. In *Coprinus*, very much enlarged structures are found in the hymenial layer that project much beyond the line of basidia. They are known as cystidia. They may be of similar shape and of different size they have abundant oily and fatty contents, and in some species, lime encrustations have been observed on the tip of these structure.

The ustilaginomycetes has a class with two orders, the Ustilaginales or smuts and Exobasidiales, important plant pathogens. The smuts of which about 1000 species are known, are biotrophic pathogens of flowering plants and cause economically important diseases of cereals. Mass of black sooty spores are formed hence the popular name smuts. The part of the plant which is commonly the most susceptible to infection and damage in flower, so smuts can cause loss of seed, such as cereal grain. The galls produced by *Ustilago maydis* on maize kernel break open to expose the sooty spores characterisitcs of smuts. The spores when mature are uninucleate and diploid are known as teliospores, teleutospores, chlamydospores or sometimes brand spores.

### *Ustilago tritici*—Loose Smut of wheat

**Habit and Occurrence:** Affected plant parts characteristically become sooty black i.e. smutty as if they have been charred or burnt up by fire. Such diseased part possess a sorus or a pustule which contains thousands of round, thick walled and dark coloured resting spores termed commonly as the brand spores or the smut spores sometimes also as chlamydospores or teliospores. They have been compared to the teliospores of the rust by being formed at the end of the life cycle on the dikaryotic sexual hyphae. Smuts are facultative saprophytes *Ustilago tritici* and *U. hordei* can be cultured on the artificial medium.

These fungi commonly parasite the flowering plants particularly the cereals belonging to Gramineae. Symptoms appear in their floral regions where an abundant powdery mass of the black brand spores replaces the normal seed or the grain. Symptoms are clear on their floral spikes turn black being filled with brand spores. In sugarcane, infected with *U. scitamineae* the entire floral axis transforms into a black whip like structure. The covering tissue of the sorus dissolves, leaving the enclosed spores of the smut parasite loose as in *U. nuda, U. avenae* and *U. tritici* the loose smut of barley, oats and wheat respectively. In the case of wheat however, the entire flower is consumed leaving ultimately the inflorescence axis bare with a few brand spores attached on its sides. In others the covering membranous tissue remains intact forming covered smut as of barley, caused by *U. hordei*. The smutted grains are often markedly hypertrophied as against the normal healthy grains *e.g.* in sawan infected by *U. paradoxa* in Bajra infected by *Tolyposporium penicillariae* and in corn, affected with *U. maydis*. In the former two cases the hypertrophy of the grains may be only 2–3 times larger while in corn smut the smutted grains may become as big as child's head. The replacement of the normal grains by the brand spore powder of the smut parasite in all these cases makes these organisms economically very important because it directly reduces the grain yield.

The general distortion and malformation of the vegetative parts of the smut infected plants, normally does not occur but sometimes the attacked plants may become stunted and dwarf as in the oats and wheats. Large mushroom like galls are produced on the stems of *Polygonum* Chinese infected by *U. treubii*. These galls are reported to be growing by the meristematic activity of the cambium, being richly supplied by the vascular supply from the host.

**Vegetative Structure:** The primary mycelium originates from the sporidia or the conidia, both of which, generally germinate in the soil, or any other substrate. In some cases, sporidia germinate on the young host tissue and produce the primary monoploid mycelium, as usual in relatively of short duration. Soon it gets diploidised to form the secondary diploid mycelium which then persists through out its life and ultimately the brand spores are formed from it. The diploidisation in most forms occurs outside the host but it may also take place. With in the host tissue as in *U. maydis*. To start with the hyphae may be present intracellularly in the host cells but later they become inter cellular with or without haustoria. The haustoria, generally are formed in the family Tilleiaceae where they may be capitate, digitate or racemosely branched. In *U. maydis* the hyphae are intercellular, penetrating the host cells and sometimes even causing death. The hyphae generally grow by keeping pace with the apical meristematic regions of the host of ultimately reach the floral spike where the brand spores are formed. The hyphae in the older regions die out gradually (Fig. 11.9).

**Asexual Reproduction:** The asexual mode of reproduction is not a common feature of occurrence but the conidia are often formed on the saprophytic stage of the smut and sometimes on conidiophores emerging out of the living host leaf surface as in *U. vuijckii*.

The conidia generally are hyaline, pyriform and either uninucleate or binucleate depending upon whether, they develop on the monoploid or the diploid hyphae respectively. In the formation of conidium the nucleus or the nuclei divide and daughter nuclei migrate into the conidium which is cut off from the parent hypha by a septum. In the latter case sometimes, only one of the two nuclei migrate into the conidium hence both the types of conidia are possible to be formed from a diploid hypha. On germination, a conidium from the monoploid hypha develops into a characteristic sexual phase of its parent but one from the diploid hypha may either develop into the dikaryotic mycelium or may produce the monokaryon hyphae of one or the other sexual phase depending upon the nature of the nucleus having migrated into it at the time of its formation. *Ustilago cardamines* cause of smut in the pods of *Cardamine bellidifolia* is reported to produce its conidia in the another of its host.

The conidia may be produced repeatedly over the same conidiophore. They germinate either by a germ tube and infect the host or may form secondary conidia by budding in the yeast like manner and thus behave as sprout cell.

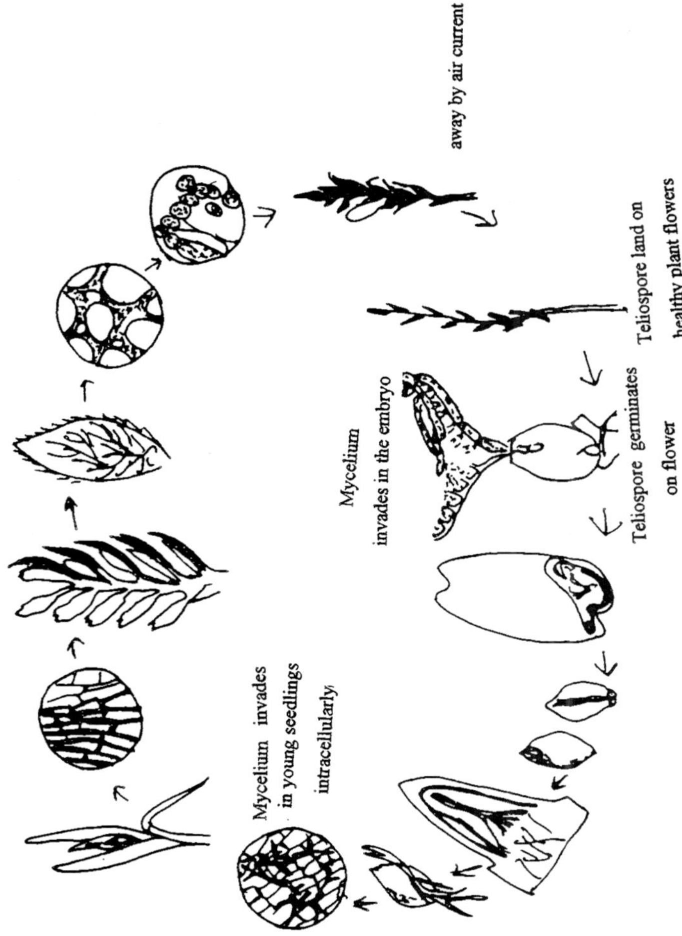


**Fig. 11.9:** Life of *Ustilago tritici* (loose smut of wheat)

**Brand spore (smut spore):** These spores develop on the dikaryotic mycelium which forms a dense tangle of hypha within the host tissue. The hyphae are composed of extremely slender but small cells, each of which carrys dikaryon. The spore mother cells enlarge their walls, gelatinise and the protoplasm becomes densely granular. Meanwhile, new walls that have to persist, are laid down beneath this gelatinous covering which gradually disappears learning the spores separate from one another, as in *Ustilago* and *Sphacelotheca* while, they are held together as spore balls in *Tolyposporium* (Fig. 11.10 A-G).

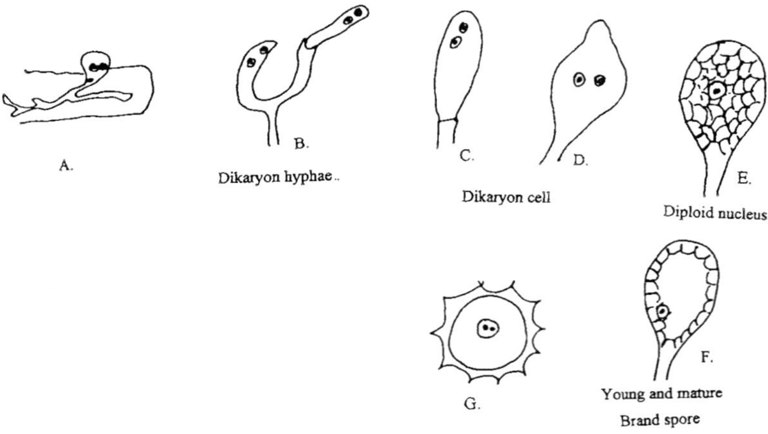


**Fig. 11.10:** Stages in development of smut spore.

The young spore is hyaline, carrying the dikaryon but as it develops, the nuclei fuse and form a single diploid nucleus. Thus the mature spore is uninucleate. It is covered by a thin endospore and an outer thick exospore which is generally dark coloured brown, black, or violet. Exospore may be smooth or echinulate. The brand spores are usually called as chlamydospores. The spores form a sort of sours with in the affected parts of the host mainly in the ovary which is completely filled by their dusty mass instead of the normal grain. These are disseminated by air and behave as resting spores. In some cases, they may germinate soon after but more often, they under go a period of rest generally until the nexι season. Average yield of viability of brand spores is between 3–10 years although the spoes of *U. hordei* are known to remain viable over 23 years.

**Germination of Brand spores:** In the presence of suitable moisture and temperature the brand spore germinate by sending out a small germ tubes called the promycelium. From the point of rupture of the exospore. Germ spores are known to occur on the spores of *U. tritici*. The diploid nucleus migrates into the promycelium and undergoes 2 divisions resulting into 4 haploid nuclei that are uniformly scatter with in the promycelium. The reduction occurs either in the first or in the second division of the diploid nuclei. The promycelium structure each uninucleate. These nuclei divide by mitosis and mean while the other remains in the promycelial cell to again divide and form another sporidium in a similar way. Large number of sporidia are produced on the promycelium. The sporidia are thin walled, oval, elliptic or sickle shaped and generally uninucleate (Fig. 11.11).

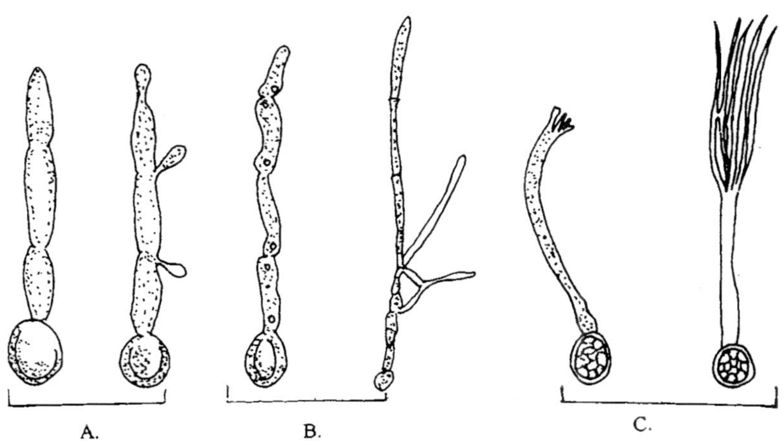


**Fig. 11.11:** Germination of smut (brand) spore

A. *Ustilago avenae*. B. *U. nuda*. C. *Tilletia carries*.

In some cases, the diploid nucleus instead of migrating into the promycelium undergoes reduction division within the spore itself as in *U. maydis* and *U. longissima*. Sporidia either germinate to develop into the monokaryon hyphae forming the primary mycelium carrying the particular sexual phase of the sporidium or they may act as sprout cells budding in yeast like manner to form several crops of secondary sporidia.

**Sexual Reproduction:** Sexual reproduction involves the formation of the dikaryotic stage. The sporidia are uninucleate and carry the haplophase of the smut while the brand spores are formed from the dikaryotic mycelium carrying the diplophase. Most of the species are heterothallic generally of the bipolar type but some like *U. maydis* are tetrapolar also the dikaryotisation may be effected by one of the following methods.

(1) **Somatogamous copulation:** Occurs between the vegetative cells or hyphae and may take place in different ways as enumerated below:–

(a) It may occur between two monoploid sprout cells belonging to the opposite sexual strains. These are formed by the breaking of the promycelium. Sprout cells grow in an yeast like manner. They may copulate either directly or by conjugation tubes and produces the dikaryon. The copulation may also occur between a sprout cell and a cell of the promycelium belonging to the different phase.

(b) It may occur between cells of two monoploid hyphae belonging to opposite strains. This is quite common method among the smuts when the monoploid hyphae develop either from the sporidia or the conidia.

(c) Sometimes in *U. hordei* and *U. levis* the promycelial cell instead of forming the sprout cells, produces the monoploid hyphae. The two sexually opposite hyphae interwine as usual; their cells copulate and develop the dikaryon.

(d) In *U. maydis* sporidium infects the young host tissues and develops the monohaploid hypha of short duration and limited growth. This does not develop further unless mating occurs by a similar hypha developed from another sporidium of the opposite sex which may also infect close by. The dikaryon thus produced, extends and persists to form the brand spores.

(e) In *U. striiformis* the brand spore germinates by producing a long and branched, monoploid mycelium in which the nuclei form pairs in certain cells whose walls thicken, the two nuclei fuse and form the spores again.

2. **Basidial Copulations:** It occurs between two cells of the same promycelium or also of another promycelium, both belonging to different sexual strains. In such cases promycelia generally do not form sporidia. The copulation may occur between two contagious cells or between repeated cells through conjugation tubes. The nucleus from one passes into the another. The diploidised all then form the dikaryotic hypha. In *U. nuda, U. tritici* and *U. avenae*. In *U. bromivora* however the diploidised promycelial cells may form the binucleate sporidia which germinate into the diploid mycelium.

3. **Sporidial Copulation:** It occurs between the two sporidia belonging to opposite phases. They may conjugate either, directly or by sending copulation tubes and thus initiate the dikaryon which later forms the diploid mycelium as in *U. scabiosae* and *U. striiformis*. Meiosis occurs in the brand spore and the usual promycelium is not formed but instead a short germ tube develops that cuts off a number of sporidia. The first formed are binucleate while subsequent ones are uninucleate. The binucleate sporidia may become 2 celled by septation, each cell uninucleate. The dikaryon is initiated by copulation between two uninucleate sporidium of opposite sexes.

4. Sometimes a sporidium may copulate with a cell of the promycelium or of the monoploid hyphae belonging to different strain.

5. **Condial Copulation:** Occurs between two conidia developed from the monoploid hyphae of opposite phases. The binucleate conidium thus formed develops the dikaryotic stage. It

occurs in those forms that produce conidia.

Smuts do not exhibit a standard type of sexual reproduction in their life cycle. They are regarded to form a degenerate group. In majority of the smut species normally a 4 celled promycelium is produced from the germinating brand spore and sporidia are formed over it in the usual way. In *U. bromivora* the promycelium is however 4 celled but several germ tubes may develop in the combination of either one celled and both 2 celled, one 2 celled and two 1 celled. In all these cases, the meiosis has occurred with in the spore itself.

**1. *Sphacelotheca:*** In this brandspores are produced singly and the functional spores are developed only in the middle of the sorus. The sori develop in the host ovary. A sterile slender column of hard tissue persists in the centre of the sorus. This is columella and is composed of host tissue as well as the fungus mycelium. The spores on germination may form or more than one septate promycelium on which the sporidia are borne terminally and laterally.

*S. sorghi* Grain smut of Jowar.

*S. cruenta* Loose smut of Jowar.

**2. *Tolyposporium:*** In which the spores form spore balls that break with difficulty. The sorus lacks the columella. The spores germinate by the production of the promycelium and the sporidia over it in the usual way.

*T. penicillariae* Smut of Bajra.

*T. ehrenbergii* Long smut of Jawar.

***Neovossia*** is characteristic in the sense that the spores do not escape out as powdery mass but instead germinate with in the host tissue and the promycelia forming tufts, emerge out through stomata. conidia on conidiophores, emerging out of the host surface, are also common. It causes smut of spinach, dahlia etc.

*N. horrida* Bunt of rice.

*N cepulae* Smut of onion.

**3. *Tilletia*** Bunt of wheat–This disease affects wheat, but not barely or oats, it is popularly called Bunt or stinking smut. Referring to the foul, fish like odour of infected grains, an odour which is unmistakable if these be rubbed between the moistened finger. In the ear the bunted grains are greyish and swollen and cause the glumes to gap apart, thus the ear looks darker and bulkier than a healthy ear. Diseased plants can sometimes be recognised before flowering time by being of deeper green colour and more luxuriant growth than normal plants. When infected grains are broken, which happens in the ordinary course during threshing, a dark brown powder is severed (Fig. 11.12B).

| | |
|---|---|
| *T. levis* | Bunt of wheat |
| *T. separata* | Bunt of rye |

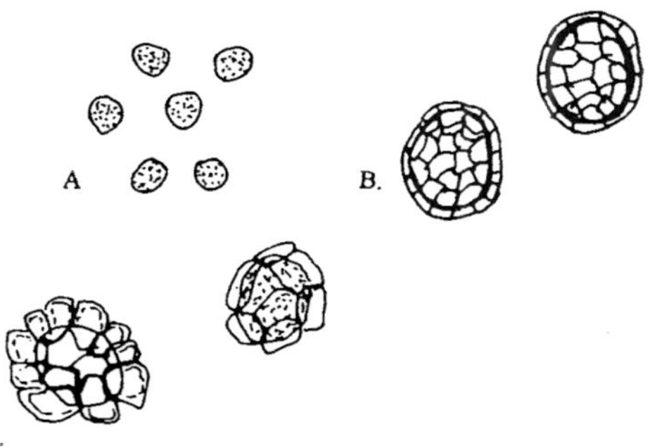


**Fig. 11.12:** Smut spores of

A. *Ustilago avenae*. B. *Tilletia caries*. C. *Urocystis agropyri*.

***Urocystis:*** It is a smut fungi and less of economic importance than *Ustilago* and *Tilletia*. Chlamydospores are formed in the leaves and stems of the host plants and are usually spore clustures consisting of one or two fertile brown cells surrounded by a number of pale sterile ones. The germination of a fertile cell is very similar to that of chlamydospore of *Tilletia* (Fig. 11.12C).

| | |
|---|---|
| *U. occulata* | Rye smut |
| *U. cepulae* | Smut of onion and Leek |
| *U. colchici* | Smut of Liliaceae. |

The Urediniomycetes are a class with a order, the Uredinales or rusts so called because of the rust coloured masses of spores produced by many species on the plants that they infect. About 6000 speices are known. All are biotrophic parasites attacking ferns, conifers and especially flowering plants.

Fungi belonging to the order uredinalcs are commonly called rusts and are characterized by the production of 4 sporidia. Over a septate, generally 4 celled, promycelium. The septa are formed transversely. Sporidia are formed over distinct sterigmata are discharged with violence. They lack a distinct sporophore. The rust occurs on obligate parasites on the ferns, conifers and flowering plants.

On the basis of nature of the teliospore structure and their mode of germination, the order uredinales is subdivided into 3 families–Pucciniaceae, Melampsoraceae and Coleosporiaceae. *Puccina* is the example of family Pucciniaceae

**1. *Puccinia:*** These forms occur as obligate parasites on their hosts forming rust coloured pustules on any or all parts of the plant above the ground. The invaded tissue are discoloured usually becoming pale green against the normal. Affected plants may even become stunted and sometimes the tissues are also hypertrophied. Parasite does not cause much serious damage to the host except, that the growth of the plants may be somewhat retarded and the grains may be reduced in size and poor quality especially in cases of infected cereals.

**Vegetative Structure:** Mycelium is septate and branched intercellular hyphae which send houstoria (bulbous, branched or knotted) into the cells for obtaining nourishment. Depending upon the infection secured in the perennial or annual plants the rust mycelium, as either perennial or short lived. Cells may be uninucleate in haploid hyphae or binucleate in dikaryotic hyphae during the life history and the number of nuclei in the haustoria may be one or two.

Generally 5 different forms and kinds of spores formed during the complete life cycle of a typical rust. These spore forms, their sori in which they are produced and the mycelium from which they originate are as follows (Fig. 11.13):

(*a*) Spermatia or Pycniospores produced in Spermogonium or Pycnium on monokaryon hyphae.

(*b*) Aeciospores produced in Aecium on dikaryon hyphae.

(*c*) Urediniospores–produced in Uredinium on dikaryon phase.

(*d*) Teliospores–produced in Telium on dikaryon hyphae. The young teliospores have the dikaryon while in the old ones the dikaryon fuses to form a single diploid nucleus.

(*e*) Basidiospores or sporidia produced on Basidium or promycelium developed on germinating teliospore.

A rust may fall in one of the following two categories, depending upon the number of spore forms and stages occurring in the life cycle:

(*i*) Macrocyclic: are the long cycle rusts which produce at least one more binucleate spore other than the teliospore e.g. *Puccinia graminis* in which all the spore forms are present.

(*ii*) Microcyclic: are short cycle rust in whose life cycle one or more spore forms are missing. Such rusts produce the teliospore as the only binucleate spore form *i.e. Puccinia malvacearum* which produces only spermogonia and telia.

In *Kunkelia* and *Endophyllum* telial stage is lacking, aeciospores germinate into a promycelium. Rust which produces different spore forms and stages to complete their life cycle are termed polymorphic.

The species of rusts which need two host plants, belonging to widely separated genera, to complete their life cycle are termed Heteroecious and the phenomenon is called as Heteroecism. *Puccinia graminis,* black stem rust of wheat, is a typical and common example of this for which two hosts are involved (*i*) *Berberis vulgaris*–a dicotyledon plant, where spermogonial and aecial stages is present (Fig. 11.13). (*ii*) *Triticum aestivum*–a monocotyledon plant–uredinial and telial stage is present. Similarly,

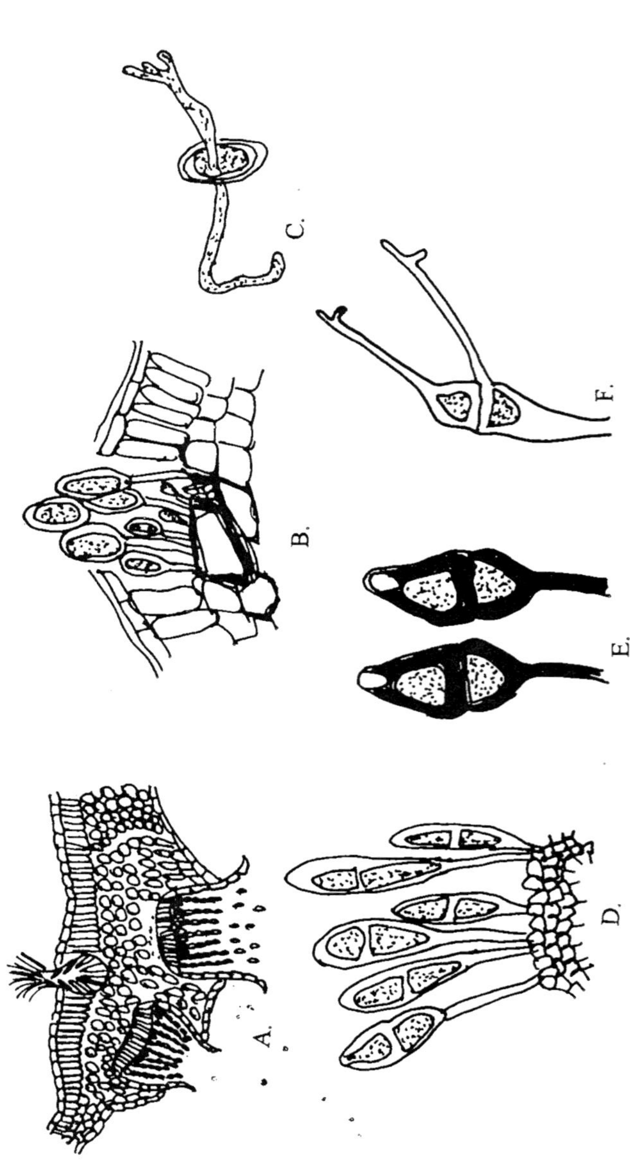


**Fig. 11.13:** *Puccinia graminis tritici*

A. Pycnidial and Aecial cup of Barberry leaf. B. Uredostage on Wheat. C. Germinating Uredospore. D. Teliospore on wheat. E. Enlarged Teliospore. F. Germinating Teliospore

*Gymnosporangium Juniperi-Virginianae* is also heteroecious by involving two hosts *i.e.* apple and juniperus.

Black stem rust of wheat--*Puccinia graminis tritici*--Heteroecious, polymorphic and macrocyclic rust form.

**I. Basidiospores or Sporidia**

They are produced on the basidium or the promycelium formed by the germinating teliospore. These are ellipsoidal or elongate with their ends either pointed or rounded. The sporidia are haploid with some what yellow contents in them. They are shot off from their sterigmata with great violence if by chance fall on the upper surface of the barberry leaf in a drop of water of thin film of water then they soon germinate. Germ tube penetrates through the cuticle and the epidermal cell wall making a very narrow pore. The hypha on the inside and outside the infected cell is comparatively thick. With in the epidermal cell the hyphae develops into a row of uninucleate cells. The hyphae soon develops between the mesophyll cells of the barberry leaf and forms an intercellular mycelium which may belong either to (+) strain or (–) strain depending upon the sexual phase of the infecting sporidium from which it has been obtained *Puccinia graminis* is a heterothallic species hence, half the crop of the sporidia is (+) and the other half is (–), the sexes having segregated at the meiosis in the teliospore during the germination.

**II. Spermatia or Pycniospores**

These are produced in the spermagonia or pycnia which develop from the monokaryon mycelium with in 4 days of the infection of the sporidia. They appear generally on the upper surface of the barberry leaf. The intercellular hyphae interwine to form a pseudoparenchymatous stroma from which develops a spermagonium as a flask shaped structure. From its base arise slender vertical parallel spermatiophore which receives the divided nucleus. It is then cut off by a transverse septum into an oval or round spermatium, which falls off. The basal mother nucleus divides repeatedly and number of spermatia are similarly cut off successively. Interspersed between the spermatiophores, they may also be present sterile paraphyses. The spermogonium opens out exteriorly by means of an apical ostiole through which emerge out a number of periphyses that are situated at the periphery of the spermogonium. Towards maturity, a fragrant sugary nectar oozes out of the ostiole which carries in it spermatic. Depending upon the sexual phase of the initial sporidium, the spermogonium and the resultant spermatic belong either to (+) or (–) strains. It has been found that the spermatic behave as the male counterpart in sexual reproduction *i.e.* diploidisation of the rust.

Spermatic are small, oval to spherical in shape, hyaline, smooth walled with very little cytoplasm and reserve food. The nucleus occupies nearly two third of its space. The insect and flies, which visit these leaves for feeding upon the sugary ooze, also help in cross fertilisation.

**III. Aeciospores**

These are produced in the aecium which develop on the lower surface on the barberry leaves from the same mycelium which has formed the spermogonia on the upper surface (Fig. 11.14). The hyphae from the base of the spermogonia extend downwards to collect either in large intercellular spaces or in the stomatal chamber. A multilayered pseudoparenchymatous mass of haploid cells is formed. This is the aecial primordium, whose sexual strain corresponds to initial infecting spordium. Diploidisation may also take place somatogamously involving fusion between two hyphal cells belonging to opposite strain lying either side by side in aecial primordium or close together between the two sori of opposite sexes. Diploid cells, thus appearing in the hyphal mass, branch out and initiate the dikaryotic hyphae each cell of which is binucleate each nucleus being of opposite strain. These cells sub epidermally arrange themselves in series to function as aecial-mother cells which elongate and conjugate divison of the dikaryon takes place. In the meantime a cell carrying the pair of divided nuclei is cut off terminally. This terminal cell enlarges and divides unequally by a transverse septum following the conjugate divison of the nuclei. The upper bigger cell is the aeciospore while the lower smaller cell forms the disjunctor, buffer or the intercalary cell. The apical portion of the mother cell again behaves similarly and thus a chain of aeciospores alternating with intercalary cells is developed, the latter dissolve and set free the aeciospores.

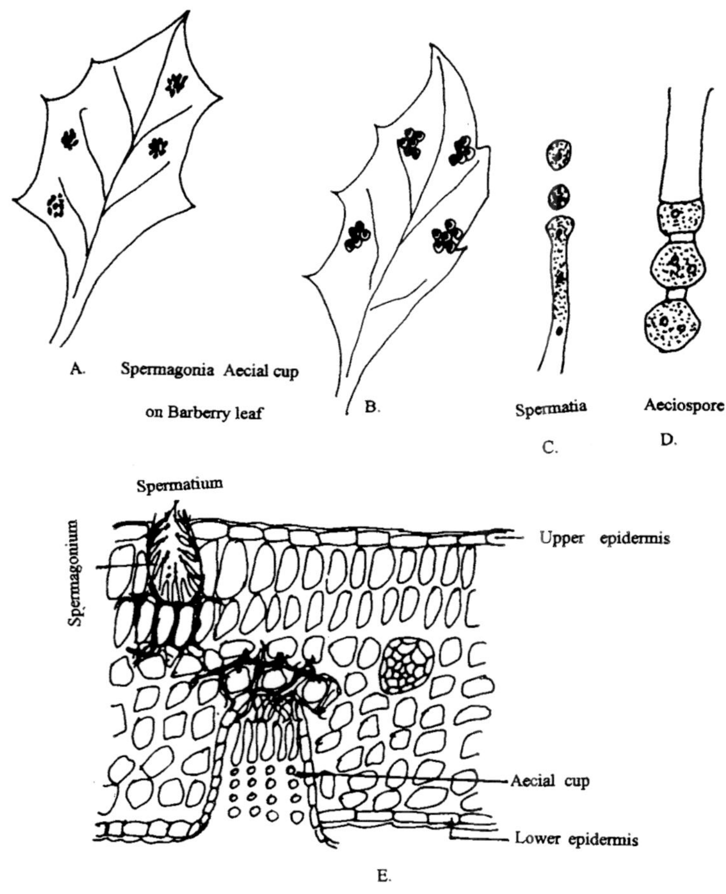


**Fig. 11.14:** T.S. of Barberry leaf through spermogonium & Aecium.

A & B. Spermatial and Aecial cup on the leaf of Barberry leaf. C. Spermatia. D. Aeciaspore. E. T.S. Barberry leaf showing spermatia and Aecial cup

The cells towards the periphery become devoid of their contents and the walls thicken due to startification. Thus a sheath is formed covering the central fertile tissue. This is the pseudoperidium or peridium which bursts through the epidermis to expose the ripe spores. The rim of the peridium breaks up recurring outwards and extending much beyond the lower surface of the infected barberry leaf giving the characteristic shape of the cluster cups as they are generally formed in groups. In a mature aecium, therefore the peripheral wall cells are monoploid while the fertile cells are diploid. Aecia are orange coloured and enclose a very large number of aeciospores within them. The aeciospores are ovate to round, single celled and binucleate. The cell contents are oily, orange, yellowish, covered by a two layered cell wall, outer layer may be echinulate. The spores at maturity is furnished with six germ pores and remain viable for long periods. They are generally disseminated by wind sometimes by rain or dew. Falling on suitable host wheat leaf, they germinate by producing a germ tube. These are incapable of infecting the barberry plant.

**IV. Urediniospores**

These are produced in the Uredinium which develops on the leaves of wheat plant generally on both surfaces as shining brown pustules which may be oval, large and elongate often coalescing to cover bigger areas. The uredinial stage is formed by the infection of the aeciospores brought from the infected barberry plants or also quite often by infection of urediniospores itself coming from the neighbouring wheat plants infected earlier.

The stout germ tube either of the aeciospore on the urediniospore swells up apically to form an elongated appresorium at the mouth of a stoma. A narrow branch from this penetrates the stomatal opening and enlarges to develop into a substomatal vesicle which has gradually received all the contents of the germtube and as multinucleate. A short hypha of dikaryon cells develops from this vesicle, in which each nucleus of the dikaryon belongs to the sexually different strains. This hypha ramifies with in the intercellular spaces of the mesophyll tissue sending haustoria and develops a limited area of infection. A layer of dikaryon cells–basal cells in formed beneath the epidermis from the sub epidermal pseudo parenchymatous stroma. The basal cells elongate vertically and divide transversely into an inferior foot cell which does not divide further Superior cell which again divides transversely into two upper larger form the urediniospore and lower elongates to form the stalk. During all this process the dikaryon keeps on dividing conjugately (Fig. 11.13B).

The urediniospores are formed and mature in 10—12 days after the initial, infection. The pressure of the developing spores bulge out the overlying epidermis which eventually breaks up to expose the urediniospores enclosed with in the uredinium. The urediniospores are inter mixed with sterile hyphae the paraphyses. Spores are yellow or orange red in colour, ovate, double walled, the outer being fleshy echinulate and binucleate. They have four germ pores round their equatorial plane. The spores at maturity get detached from their stalks and disseminated by wind. These when fall on fresh wheat leaves, produce the infection and the uredinium in a similar way. Thus a fresh crop of urediniospores is ready with in 10–12 days. This behaviour of repeating their spore form is the short cycle of *Puccinia* and these spores are comparable to conidia.

It often produces thick and strong walled urediniospores with persistent stalk. These can over summer or over winter and are termed as amphispores.

**V. Teliospores**

They are produced in the telium which generally develop on the stems of wheat plant but also occur on leaves, as elongated black either, separate or coalescing pustules (Fig. 11.13D-E). This characterises the common name of the disease as black stem rust of wheat. The telia are produced towards the end of season when the photosynthetic activity of the host plant also become low. Teliospores appear mixed with urediniospores in the same pustule. Development of these spores is similar to urediniospores. But in some case upper superior cells divides further to form a two celled teliospore each cell carrying the dikaryon. Sometimes single celled teliospores are formed among the normal bicelled ones. These are termed merospores. The cell wall become thick with age and the dikaryons of each cell fuses to form a diploid nucleus. Thus a mature teliospore is 2 celled, uninucleate structure with one germ pore on each cell of the spore. These spores are dark brown or even black and survive for long periods.

On germination in favourable conditions, either one or both the cells of the teliospore send out a promycelium emerging through germ pore. The diploid nucleus may migrate into the promycelium and divide or the divison may take place with in the spore cell and then the, divided nuclei migrate into the promycelium. Two divisons result into 4 nuclei which are haploid. The reduction taking place generally in the first divison. The promycelium by three septa divides into 4 cells each of which forms a lateral sterigma over which a basidiospore is borne. The nucleus of the cell migrates into the sporidium through the sterigmata. The sporidia are uninucleate, half of the crop on each promycelium belongs to one strain, and other half to another (–) strain, the regeneration taking place at the reduction divison. The basidiospores are shot off explosively and develop the spermogonium. If by chance they fall on the barberry leaf. Thus life cycle is completed.

**Reoccurrence and Epidemics of Rust in India**

This work was carried out by Prof. K.C. Mehta (1929, 1940, 1952). In the plains of India, the cereals are sown in Octobers–November and harvested in March-April or May. Weather conditions as well as the age of host are quite favourable for infection of the crops Initial out breaks are delayed by 2–3 months from the time of sowing of the wheat crops in plains. Chances are very less for the survival of uredospores of any three rust of wheat from the previous crops in the plains due to intense summer heat. Prof. K.C. Mehta discussed the role of Barberries as a collateral host. He reported the survival of the uredo stage in all three rusts of wheat out of season during summer in the hills. He has also dissertated on the out break of rust on new crops sown earlier in the year at the foot hills. A detailed comparison shows that there is heavier infection at the foot hills than in neighbouring plains and also that the periods of rust appearance at the foot hills and the plain in different parts of the country are not same.

Mehta (1940, 1952) safely concluded that the rust inoculum is reintroduced into the plains of India year after year from some other source and this source in all probability, lies in the hills where all the three rusts are able to over summer in the uredo stage (altitudes 4000–8000 ft). On account of milder summer and requisite humidity, the rust over summers in the uredo stage on volunter wheat, stubbles etc. and on the regular crops in the Nilgiris and Pulney hills and the submountains regions in the Himalayas in the north. These were regarded as the chief foci of infection, but regular surveys conducted in 1953–54 and 1954–55 have shown that the Panchgani-Mahabaleshwar range in the Western Ghats is also a focus of infection. According to Prof. K.C. Mehta (1929, 40, 52) primary inoculum of black rust comes from the north as well as South. Joshi et al., (1974) suggest on the basis of data collected by survey teams and a large number of cooperators that South India gets black rust infection much earlier than the North and that certain regions of Central India may serve as reserviors of Inculum for the main wheat belt in the North. Although the black rust can survive in the northern hills throughout the summer months a number of factors seem to indicate that this inoculum does not playing any significant part in the spread of the rust to the plain in the North. Main reason for this is that the average minimum temperature for the months of December, January and February in North India ranges from 17–11 and 9–13°C respectively. The incubation period of the rust in these temperature will be as long as 40–50 days in the months of January and February. The inoculum builds up in the South and its dissemination towards the North appears to be the principal source of black rust inoculum to the Northern wheat crop. Karnataka, M.P. and Maharashtra are the states which serve as the chief areas of multiplication of black rust.

Black rust of wheat–*Puccinia graminis tritici*
Brown rust of wheat–*P. recondita*
Yellow rust of wheat–*P. striiformis.*
Rust of Linseed–*Melampsora lini.*
Rust of Beans–*Uromyces appendiculatus.*
Rust of Pea–*U. fabae.*

**2.** ***Melampsora:*** It causes rust of Linseed (*Linum usitatissimum*). Linseed rust is caused by *Melampsora lini.* Telial stage is present on linseed. Telitospores are elongated, compactly arranged in subepidermal region. They are sessile and red-brown in colour. Uredial stage is present on *Euphorbia geniculata* (Fig. 11.15A-C).

**3.** ***Uromyces:*** Gram rust and Pea rust is caused by *Uromyces.* It is an autoecious rust producing all the five spore forms as in *Puccinia,* but the teliospores are single celled with one apical germ pore. It is a common disease in India. The annual reoccurrence is by perennating teliospores in the soil which germinate and initiate the infection next reason (Fig. 11.16).

**4.** ***Ravenelia:*** It causes rust on the *Albizzia lebbeck.* It is an autoecious rust produced teliospores, which are one celled, adhering firmly to form an umbrella like head of one layer of fertile spores subtended by colourless cysts which probably represent sterile teliospores. The head is supported by a central stalk that is several cell thick. This rust attacks a large number of host belonging to Leguminosae (Fig. 11.16D).

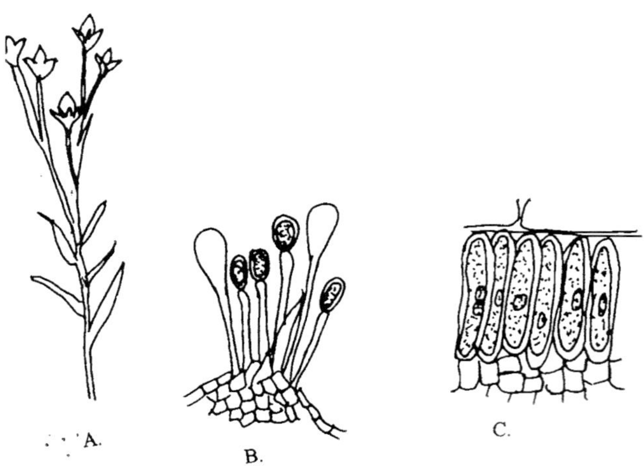


**Fig.11.15:** *Melampsora lini* (Linseed rust)

A. Symptoms. B. Uredospore. C. Teliospores.

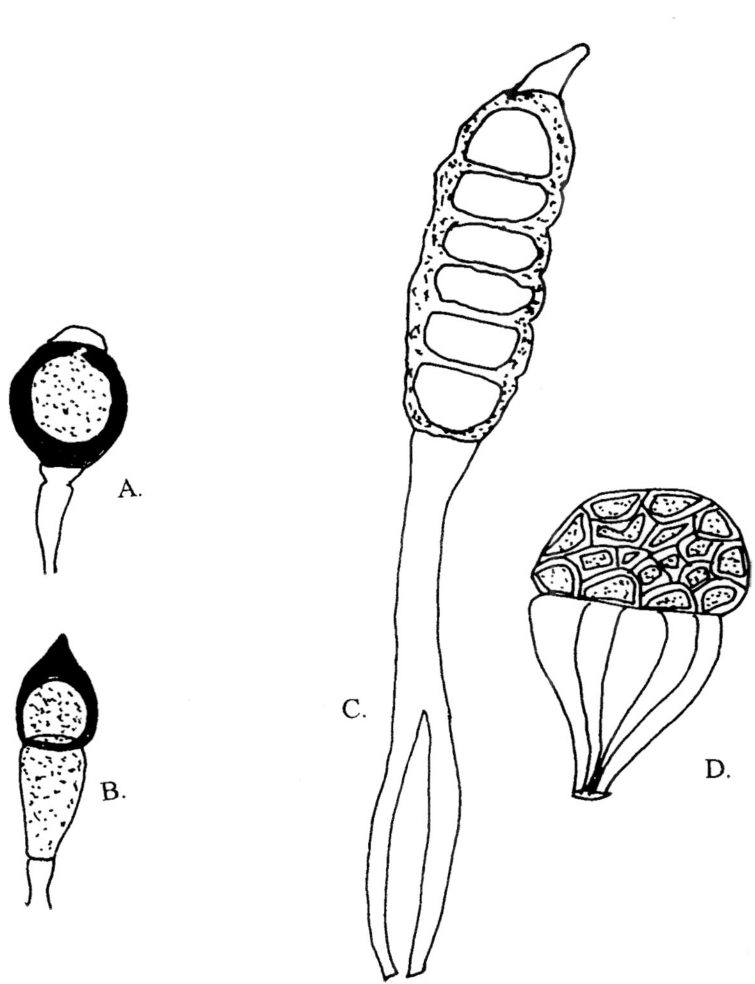


**Fig. 11.16:** Teliospores of

A. *Uromyces*. B. *Puccinia*. C. *Phragmidium*. D. *Ravenellia*

**5. *Phragmidium:*** It causes rust on *Rubus*. It is autoecious rust characterised by producing many celled teliospore borne on long stalks which are enlarged towards the base. The spermogonia are subcuticular, flat and undefined and the aecia are also not covered by a pseudoperidium. They are flat, naked or covered by paraphyses, spreading laterally and assuming an indefinite form (Fig. 11.16C).

**Hymenomycetes**

It is probably the largest group of the Basidiomycetes embracing 11,000 species. These mainly occur as saprophytes but parasitic also. The origin and development of the hymenium may be of one of the following types:

**Gymnocarpic:** In which hymenium remains exposed throughout the development. The primordium of the basidiocarp is cylindrical with tapering apex. The apical region broadens, developing into the fertile head—the pileus on the lower side of which formed of the hymenium bearing gills. The basal portion enlarged as the sheath *e.g. Omphalia*.

**Hemi-Pseudo Angiocarpic:** In this type of development, hymenium is enclosed falsely for some part of the development. The edges of the broadening pileus curve down to come in loose contact with stipe. Thus a closed cavity is formed on the roof of which the hymenium is present, which is exposed at maturity when the connecting hyphal tissue between the pileus edge and the stipe breaks away e.g. Lentinus.

**Angiocarpic:** In which the hymenium develops with in the tissue of the pileus. An internal circular chamber or cavity develops with in the pileus into which the radiating gill bearing hymenium is formed. These are exposed to shed off their spores only at maturity when the expansion and flattening of the pileus breaks open the tissue that connects it with the stipe *Agaricus*.

Hymenomycetes are divided into many order. Among that two are discussed here.

(a) **Agaricales**—*gill fungi*

(b) **Aphyllophorales**—*bracket fungi.*

**Agaricales.**

Forms belonging to this order are commonly termed as gill fungi comprising the mushrooms and toadstools that occur mainly as saprophytes growing on grounds in the fields and lawns, leave bark, wood logs, manure heaps etc. They form a common sight in such habitats immediately after a shower of rain. Some form like *Asterophora* and *Volvaria* occur as parasites over other *Agricus*. In *Armillaria mellea* rhizomorphs may be phosphorescent in dark thus exhibiting bioluminescence. Its slender branches penetrate the roots and the collar region of the host tree destroy their cortex causing the 'white rot' and ultimate death of the tree. The oak tree are very susceptible to infection by *Armillaria. Schizophyllum commence* is a weak wound parasite.

The Agaricales form a cosmopolitan group and are important because they aid in the decomposition of the dead plant tissues in the soil and they also occur as a mycorrhiza.

The primary mycelium developing from the basidiospores soon get diploidised forming the secondary mycelium which perennates in the soil or substratum. The hyphal knots towards the periphery form the sporophores which during favourable conditions, develop and grow very rapidly appearing above ground almost over night. The sporophores are thus formed either in a ring or in incomplete arcs. These generally have bright and beautiful colours hence they have been termed as fairy-rings, marking the paths of the **dancing fairies**. *Marasmius oreades* the fairy ring mushroom is perhaps the best known of a number of mushrooms that form rings. Inside the ring of mushrooms there is usually a distinct zone of grass that is noticably greener than the grass. The greener color is due to the nitrogenous substances that become available to the grass as the older hyphae of the fungus mycelium die and disintegrate.

An interesting phenomenon exhibited by a number of agaries in bioluminiscence. The reproductive structure produced by species of *Mycena lux coeli* and hyphae give off visible light causing them to glow in the dark. This phosphorescent glow has long fascinated and frightened humans. This phenomenon commonly called fox-fire has contributed significantly to fungal folk love as well as to the overall mystique of fungi. A number of Agaries found in North America are bioluminescent including *Armillaria mellea, Panellus stypticus* etc. In *Armillaria* only the mycelium and rhizomorphs are luminescent. By this process inscets are attracted.

The asexual mode of reproduction is not of common occurrence but in some *Agaricus e.g. Coprinus* and *Collybia*, the oidia are formed both on the monoploid and diploid hyphae. The oidia may also have a sexual function in bringing about diploidisation. Chlamydospores are of quite common occurrence in the group.

The sexual reproduction is mainly somatogamous where fusion occurs between two cells of the monoploid hyphae belonging to opposite strains. The resulting diploid mycelium initiates development of the sporophore. The basidiocarps are generally fleshy conditions, becoming leathry, corky or even woody. These exhibit a characteristic form *i.e.* they have a central stalk—the stipe which may rarely be lateral or even absent in the *Schizophyllum, Pleurotus*.The stalks supports an open umbrella like cap—the pileus which is the fertile head. The pileus may be of various shapes and size bearing thin lamellae—the gills which hang downwardly on its under surface. The gills are lined with continuous hymenium comprising the basidia, basidiospores and the paraphyses. The order Agaricales has been subdivided into 5 families—Boletaceae, Paxillaceae, Russulaceae, Hygrophoraceae and Agaricaceae. A detailed description of the life history of *Agaricus:*—

### *Agaricus*

**Habit and occurrence:** *Agaricus* occurs as a saprophyte growing on the ground in the fields and lawns, wood logs and manure piles. *A. campestris* is the most common form of mushroom which, being edible, is also cultivated for commerce. These usually appear immediately after a shower of rain in July-August. They are generally stalked forms which is attached centrally to the pileus. The gills have vertically down on undersurface of the pileus. The pileus size, shape and structure exhibits great variation within the genus.

**Vegetative structure:** The primary mycelium develop from the uninucleate basidiospores. The cells of the hyphae are also uninucleate and the mycelium is short lived. It soon gets diploidised to form the secondary binucleate condition of the mycelium which perennates in the substratum and in the soil for many, years. The 'secondary mycelium' later develops a 'tertiary mycelium' that is composed of specialised secondary hyphae destined to make the complex tissues of the fruit body (Sporophore, sporocarp and basidiocarp) bearing basidia etc.

**Asexual Reproduction:** Asexual Reproduction is not common in *Agaricus*. Chlamydospores are present, they may be terminal or intercalary in position. The walls of the chlamydospores may be smooth or verrucose. Oidia may be formed under certain conditions which are also known to have a sexual function in the diploidisation.

**Sexual Reproduction:** Sexual reproduction is mainly somatogamous or pseudogamous. It involves the diploidisation process and occurs generally by the fusion between two cells of the monoploid hyphae belonging to opposite strains. Diploidisation can be achieved by fusion between an oidium and a hyphal monoploid cell. In *Agaricus campestris* hyphae is homothallic, while in other forms it is heterothallic. Heterothallic species may be bipolar as well as tetrapolar.

**Development of the Basidiocarp:** Development of the basidiocarp generally initiated over the dikaryotic mycelium. Small hyphal knots develop over the subterranean rhizomorphic perennating mycelium. These knots are ovoid to globose and form the button stage or the basidiocarp primordium. Each of the primordium in a solid mass of interwoven hyphae, towards the periphery it is less dense. This hyphae elongates and develops rapidly. Hymenial primordium is comprising of parallel vertical hyphae which form an internal ring in its transverse plane (Fig. 11.18A-F). The portion above this is the future pileus and the basal part of is to become its stipe. Following this, annular cavity develops internally between the vertical and the underlying hyphae on the lower face of the hymenial primordium. This is the prelamellar chamber whose upper surface becomes deeply concave and is lined with alternating radial bands of slow and fast dividing cells, the latter forming gill primordium which soon develop into lamella—the gill that hangs downwardly into the prelamellar chamber. Later on gill develops the hymenium surface externally. The pileus, has a tendency to greatly increase in the radial interspace between the gills. These spaces gets filled up with secondary and tertiary lamellae developing either as branches and Y like forkings of the primary series. Lamellae may split lengthwise.

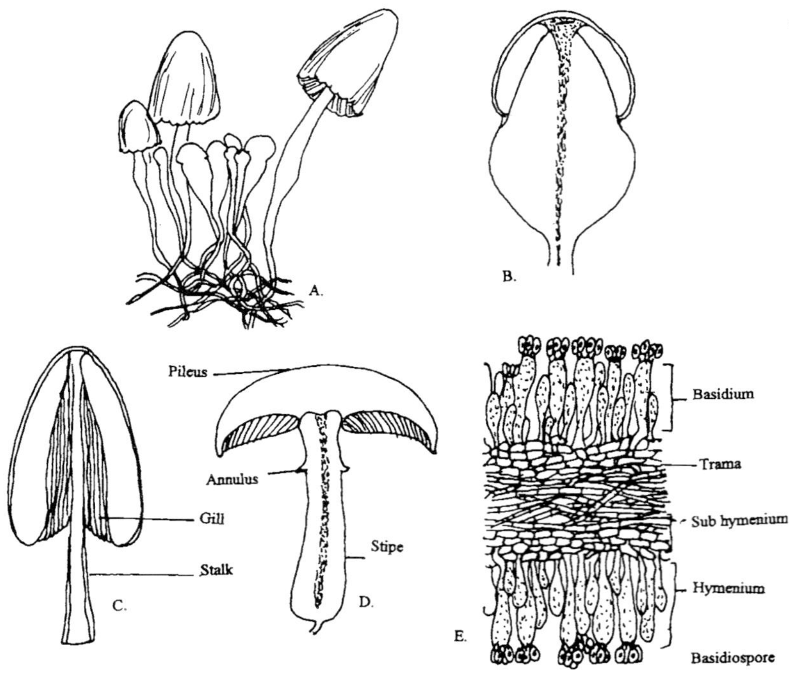


**Fig. 11.17:** Different stages of *Agaricus*

A. Aerial Mushroom. B. Button stage. C. Mature stage. D. Typical Mushroom. E. Section of gill

The pileus edge in young primordia is joined below with the stipe by a connecting hyphal tissue. This is termed as veil or the velum. This connection ruptures, when the pileus starts enlarging and opening out but in most cases, its remnants persist as a ring the annulus round the upper portion of the stipe (Fig. 11.18)

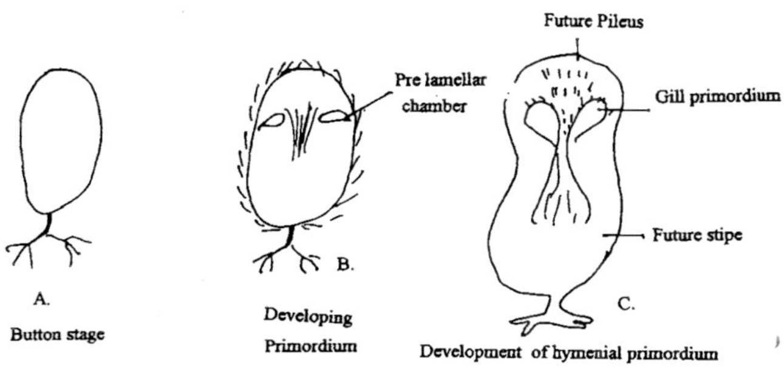


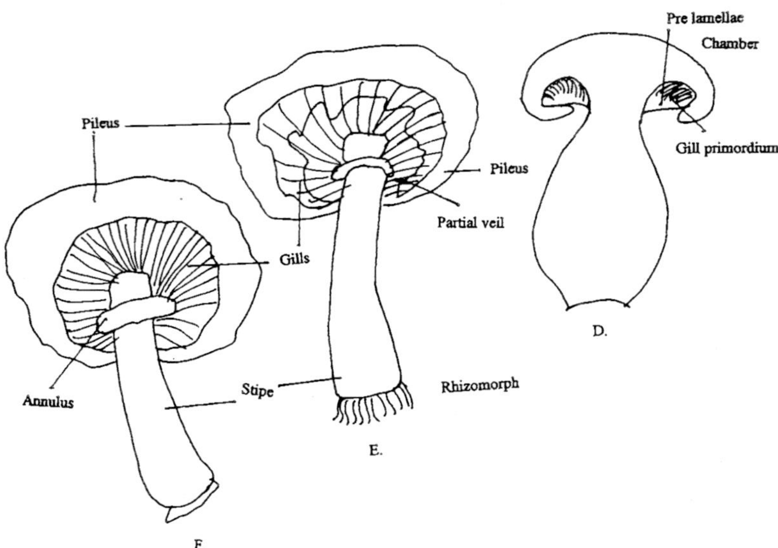


**Fig. 11.18:** Basidiocarp development of *Agaricus* (A-F)

The pileus expands out in an open umbrella like manner with numerous gills attached on its lower surface seemingly radiating out from the stipe. A fully mature pileus may bear 300-600 gills. The rapid development of the basidiocarp from the buttom of its 'umbrealla' stage is mainly due to the elongation of the existing cells in which not many new cells are added (Fig. 11.18).

In some cases, after the partial veil is torn off, a thin sheet of cobwebby mycelium is left hanging down the pileus edge. This is termed as cortina meaning a curtain as in *Cortinarius*.

A section through mature gill reveals that it is composed of a central core—the trama comprising parallel or loosely interwoven hyphae which totally give out oblique branches running out towards to form a compact layer—the subhymenium. Branches emerge out almost at right angles to the subhymenium and develop a palisade like layer comprising basidia and the paraphyses. Both these asci from the dikaryotic hyphae and form the hymenium layer (Fig. 11.17 E).

Each basidium is a cylindrical structure with tapering base and an obtusely rounded apex. It may be clavate or even globose. The young basidium carries the dikaryon, the two nuclei, fuse and soon under go two successive divisions—one of them being reduction, forming four haploid nuclei arranged at the top of the basidium. Four small horn like sterigmata develop apically on the basidium, each of which buds out a basidiospore at its tip. The wall of the basidiospore is delicate, enclosing a dense cytoplasm with in it. A large vacoule develops in the basidium due to which the cytoplasm and the nuclei migrate into the budding basidiospore whose wall thickened. Billions of spores may be shed off with in a few hours only. They fall vertically down. This behaviour of the spore fall can be utilised to study the arrangement and nature of the gills on the pileus, by making spore prints. The mature umberalla part of the mushroom is cut off from the apex of the stipe and placed for an hour or so over a piece of paper of the contrasting colour than that of the *Agricus e.g.* white paper for dark coloured sporophores and black paper for white spored forms. As the spores fall on the paper, the interspaces of the gills become accordingly, marked and the manner of the branching is distinguished. Most of the mushrooms, at maturity undergo autodigestion in which, the gills deliquesce due to the enzymatic autolysis of the exhausted lamellae from the pileus edge of the stipe.

The hymenium surface covering the gills may, in most cases, have only the basidia and the paraphyses or may also have charger basidia like strutures called cystidia. Cystidia generally project much beyond the hymenial surface. They are sessile, devoid of sterigmata much elongated and broad, pointed, knobbed or forked apically. The upper portion is often covered with crystals and cystidia. They also have abundant oily and fatty contents with their walls being occasionally thickened. Being commonly present in between the two adjacent gills and strectching is to keep the gills apart especially at the time of autolysis and allow the spore fall to continue for long.

***Polyporus***

**Habit and Occurrence:** It causes severe wood rotting. It commonly occurs as a sapraphyte on the ground or on dead logs. It mainly parasites on the forest and shade trees. By the persisting mycelium it spreads mainly in the sap and heart wood. Basidiocarps are generally, leathery and tough forming bracket like structures with or without a distinct stalk.

**Vegetative structure:** Mycelium is slender and much branched hyphae composed of long cells. The primary monoploid hyphae develop from the uninucleate basidiospores are short lived and soon get diploidised to produce the perennating secondary dikaryotic mycelium over which is initiated the development of the sporophore.

Mycelium initiate its life either in the soil from where it enters the host root or it enters through a wound in the host tissue; quickly ramifying in its phloem and xylem tissues. The hyphae developed in the thin walled cells below the bark where they collect to form a white solid paper like crust. This is the seat of basidiocarp development.

**Asexual Reproduction:** This type of reproduction is not very common in *Polyporus*.

**Sexual Reproduction:** During sexual reproduction, fusion between the adjacent cells takes place, which belong to different mycelial strains. Most of the species are heterothallic.

**Development of the Basidiocarp:** Basidiocarp development initials from the dikaryotic mycelium. It appears as a more or less spherical knob of hyphae on the mycelial mat that stretches just below the bark of the tree or log of wood. It gradually emerges out and may be shortly stipitate, when young. The upper surface is much infolded. During development upper surface becomes flat and smooth while, the lower attains a rounded or remiform shape. It is spongy and covered by a downy layer of fine mycelium. The pores of varying shapes and size, depending on the species, develop on the lower surface which gradually becomes hard. The development of the spores is centrifugal *i.e.* younges ones towards the periphery.

A longitudinal section of the young sporophore reveals the upper surface to be composed of parallel running thickened hyphae that form a rough layer of oval cells on the free surface. It is composed of hyphae enclosing large irregular spaces between them. These hyphae towards the lower external surface, dissolve at the margin and open out as pores or tubes. The body space between the closely adjacent tubes is composed of loosely interwoven longitudinal hyphae which form the trama or the pseudoparenchymatous stroma. These develop short branches at right angles to the tubes length forming the hymenium that lives the inner surface. Sometimes a distinct layer of obliquely running and compact hyphae occurs between the hymenium and trama this is the subhymenium. The hymenium comprises the basidia fuses with the young basidium and divides generally in four haploid nuclei and sometimes by meiosis into eight. Four small, oval, nucleate basidiospores develop over short sterigmatas which are later discharged with pore space drop down and are disseminated by wind. The spore discharge continues for weeks and even for months during which time, billions of spores may be formed. The hymenium remains confined merely to the riges of pores and is usually not formed on their edges (Fig. 11.19 A-B).

***Ganoderma:*** It is an annual form whose sphorophore have varnish like coating over the stipe as well as over the hard crust of the pileus. It occurs on fallen trees: old stumps or on coniferous and deciduous trees (Fig. 11.19C).

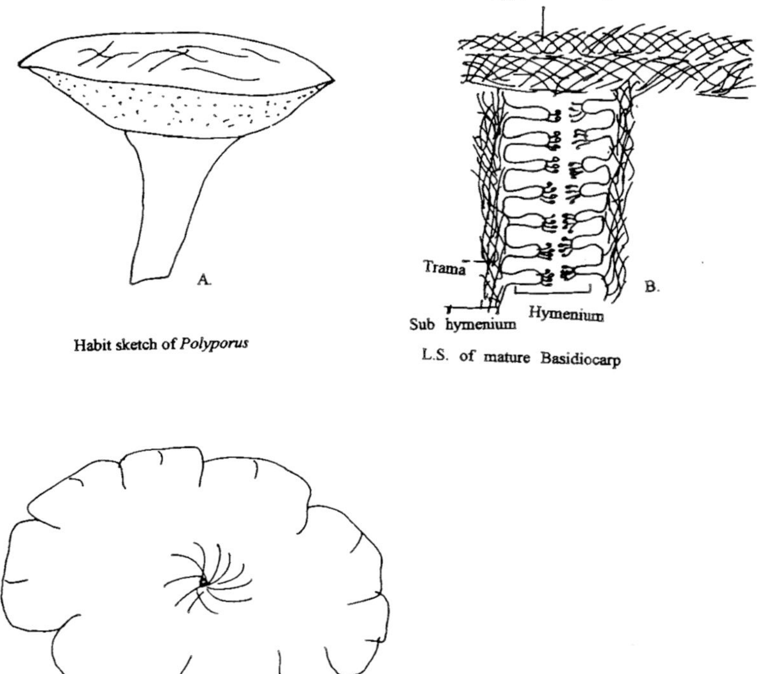


**Fig. 11.19:** A Habit sketch of *Polyporus* B. *L.S.* of mature Basidiocarp C. Habit sketch of *Ganoderma*

# 12. DEUTEROMYCETES

There are many species of fungi of which perfect stage is not known and have lost the power to produce a perfect stage is known as **Fungi Imperfecti**. Distribution of thousand species of fungi imperfecti into genera, families and orders most necessarily be based upon the vegetative and asexual reproductive stages. Imperfect fungi can be sub-divided into groups of species corelated with perfect stages. Many imperfect fungi have several different spore forms makes their recognition as definite species difficult, even where the perfect stages are not discovered. There are indications to prove that the bulk of the forms described under Deuteromycetes, represent the conidial stages of Ascomycetes and a few of Basidiomycetes. Members belonging to this class of fungi have septate mycelium and reproduce by means of asexual spores only. Fungi that live in wet habitats frequently produce their spores on hyphae that are completely submerged *i.e. Tetracladium, Tetrachatum, Lenulospora* etc. The spores are wholly slender and with long branches which give them a great ability to float.

Many species and genera are usually divided into four form orders as follows.

1. Sphaeropsidales
2. Melanconiales
3. Moniliales
4. Mycelia sterila

**Shpaeropsidales:** Conidia produced with in the pycnidia or modifications of such structures. A pycnidium is a perithecium like structure and may be complete, like the perithecium of the Sphaeriales and Hypocreales. It may open by a longitudinal slit as in the apothecia of the hypocreales or may be closed at first and finally open into a cup or saucer shaped structure, like a miniature apothecium.

**Melanconiales:** Conidia produced singly or in chains. Often surrounded by a gummy mass from conidiophores packed closely in a usually subepidermal or subcortical layer the **acervulus**.

**Moniliales:** Conidia formed on conidiophores which are separate at least at their apical portions, the vegetative mycelium breaking up into conidia. Conidiophores may be simple or branched, short or long. Similar to the vegetative mycelium or very distinct from it, but are never enclosed with in a **pycnidium** nor packed laterally into a subepidermal or subcortical acervulus. They are always almost external at the time of conidium production.

**Mycelia sterila:** Imperfect fungi which lack all conidial formation, which produce sclerotia, rhizomorphs and various other forms of mycelium without spores.

**Habitat of Deuteromycetes:** Conidial fungi are found in the same habitats that their sexual relatives occupy. They generally occurs as saprophytes in soil, on decaying organic matter etc., but quite a large number of forms are capable of becoming parasites, sometimes causes severe disease in plants, animals and man. Leaf spots blight, blotch, wilts, anthracnose, scab, root and fruit rots are some of the commonly occurring disease in a large variety of plants, both wild and cultivated. In animals and man the diseases caused by the deuteromycetous fungi are meningitis, candidiasis of lungs, skin, nails, dermatomycosis like ringworm, athlets foot etc.

**Vegetative structure:** Deuteromycetes typically produce well developed, branched and septate hyphae with cells usually multinucleate. The cross walls between the cells are perforated as in Ascomycetes, through which protoplasmic strands are connected thus, maintaining the continuity of protoplasm. There may be a dolipore septa in basidiomycetous asexual fungi. Some conidial fungi produce appresoria, haustoria, nematode traps and lichen thalli.

**Structure associated with asexual reproduction:** Asexual reproduction typically takes place by conidia formation, which are present on the conidiophores. Conidia may be present on the complex and specialised like structure like synnemata, sporodochia, pycnidia and acervuli. A brief description of these structure have given below.

**Conidia:** A conidium is a non motile asexual spore formed at tip or side of a sporogenous cell. It should be emphasized that conidia are not produced as a result of progressive cleavage of cytoplasm. Conidia germinates by germ tube formation to produce extensive mycelium. This process is known as macrocylic condiation. Some fungi may germinates by producing conidia directly from an ascospore or conidium. This is the microcyclic conidiation.

In yeast, conidia accumulate after formation in a wet drop let, which may be water or animal dispersed, other are produced in dry masses and the small spores are dispersed efficiently by wind. The conidiophores, when grouped together, may be arranged in a brush like fashion forming the 'coremium', in the form of hemispherical cushion known as the 'sporodochium'. Different types of conidia produced by deuteromycetes. Their shape may be spherical, ovoid, elongate, cylindric, thread like spirally curved or branched. They may be one to many celled, with either transverse septa or both transverse and longitudinal septa condition is known as muriform. Conidia may be hyaline or coloured in shades of yellow, pink, green brown or black.

**Synnema:** Synnema consists of a group of conidiophores often united at the base and part of the way of their length conidia may be formed along the length of the synnema or only at its apex. Conidiophores comprising a synnema often are branched at the top, with the conidia arising from conidiogenous cells at the tips of the numerous branches. In some synnemata stalk like portion is longer in comparison to the branched top and the entire structure resemble a long handled feather duster.

**Pycnidium:** Pycnidum is a globose of flask shaped pseudoparenchymatous structure that is lined on the inside with conidiophores. In external appearance some pycnidia resemble perithecia. Pycnidia may be completely close or may have an opening, they may be provided with a small papilla or with long neck leading to the opening, they vary greatly in size, shape and colour. They may be superficial or sunken in substrate, they may be formed by loose mycelium or definitely stromatic.

**Acervulus:** It is a flat or saucer shaped bed of short conidiophores growing side by side and arising from a more or less stromatic mass of hyphae. In nature, acervuli typically are produced in plant tissue subepidermally, or subcuticularly and break through the plant material to become erumpent. In some acervuli certain erect, dark and stiff bristle like structures termed setae are observed occurring mixed with the conidiophores.

Other asexual propagules are sclerotia, bubils and chlamydospores. Chlamydospores are thick walled cells that develop from single hyphal compartments. They usually are more rounded in shape than the typical hyphal cell and may be terminally or intercalary. Bubils are masses of psudoparenchymata that are not internally differentiated and often are hyaline. **Sclerotia** are masses of cells that form a hard, rounded structure with a differentiated ring in or on a host. They may include host tissue. Stromata are similar to sclerotia as they also are hard masses of many cells that may include host tissue but they are irregular is shape and not rounded.

**Melanconiales**

***Colletotrichum:*** It is a facultative parasite occurring usually as a parasitic form causing the **anthracnose** disease on leaves, young twigs and fruits of several plant species. The chief visual symptoms that are produced on most of the affected part are the development of pinkish or dark brown lesions resulting ultimately, in the formation of deep cankers. The red rot of sugarcane is one of the most commonly occurring diseases caused by a species of this fungus namely *C. falcatum*.

**Vegetative structure:** In culture mycelium comprises profusely branched and septate hyphae that are hyaline to start, becoming dark at maturity. Often with age, the cell contents become denser and the walls turn darker become sometimes thicker also. The thick walled hyphae, occassionally, form dark green or black stromata by becoming closely intertwined with one another. In the host, the initial mycelium develops with in the epidermal cells after it has penetrated the cuticle and manifests in the subcuticular region. It later on, develops into intercellular as well as intracellular hyphae all over the underlying and adjacent tissues (Fig. 12.1 A).

**Reproduction:** In *Colletotrichum*, only method of reproduction is by formation of asexual bodies *i.e.* conidia, although, the perfect stage of some of the forms of *Colletotrichum* has been found to be that of *Glomerella cingulata* belonging to the ascomycetes.

In culture, conidia are formed singly at hyphal ends which at maturity shift to one side allowing the main hypha to grow further and repeat the process. On the host, conidia are formed in the acervuli developing over the stromatic hyphal masses that are formed under the cuticle and are exposed at maturity by its rupture. Characteristic feature of this genus that differentiates it from other allied genera is the occurrence of a number of straight stiff, hyaline or coloured sometimes thick walled 1–3 septate, bristle like structure termed setae around the acervulus (Fig. 12.1 B).

Conidia are formed acrogenously over short conidiophores appearing all over the acervulus. They are aseptate, fusoid. Somewhat curved or sickle shaped, often cylindrical, hyaline or slightly pink coloured with rounded ends. A clear vacoule like globule occurrs centrally. When formed copiously, they occur in gelatinous masses. The conidia germinate by producing 1–4 germ tubes and form the new mycelium.

Old mycelium is thick walled, dark brown irregular or angular cells develop with one or more large oil drops within them. They are either formed terminally or in intercalary position anywhere on the hyphae. They get detached and are capable of germination after long period of rest, behave like chlamydospores.

*Colletotrichum* often cause quite destructive diseases of plants over which it occurs as a parasite producing characteristic anthracnose symptoms generally on the aerial parts of the infected host plant. One the most well known species of *Colletotrichum* is *C. falcatum*, the causal organism of the red rot disease of surgcane.

Some of the other commonly occurring species of the genus are *C. gloeosporioides*, parasitising and becoming destructive on species of citrus; *C. lindemuthianum* causing of bean anthracnose, *C. circinans* which produces black smudgy smptoms on the onion scales.

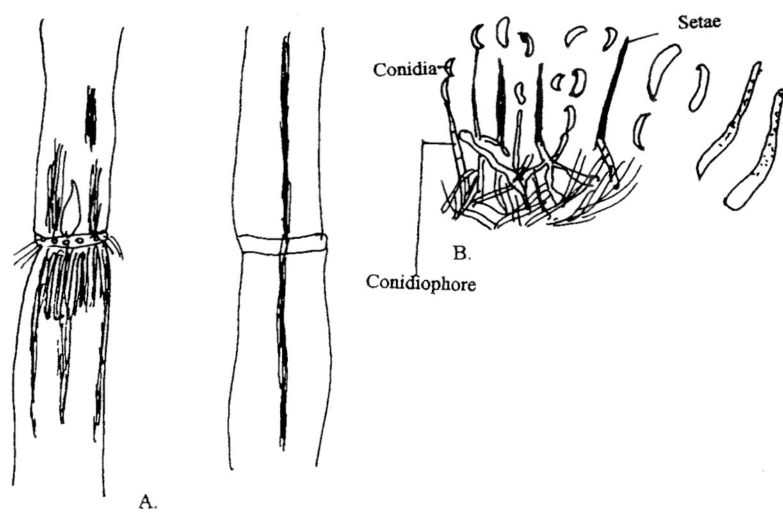


**Fig. 12.1:**

A. Symptoms of the sugar cane. B. Acervuli of *Colletotrichum*

*Colletotrichum lindemuthianum:* Bean Pod Scab. Dwarf and runner beans are attacked by this fungus. The disease appears on the pods, which are the parts more usually attacked, first as small, dark spots. These increase in size, often running together and forming irregular sunken patches of dead tissue, the border of the patch is often tinged with red, while the centre is of rust coloured. In severe cases the mycelium extends from the pod inwards to the seeds, which are thus spotted, while the pod is itself is distorted. Similar spots may be found on stem and leaves. The mycelium breaks through the epidermis and appears as exposed, pink sporiferous cushions or stromata, on which are simple conidiophores bearing conidia singly. Along with the conidiophores or confined. Sometimes to the border of the stroma, are long, blackish bristles, these are characteristic of the genus *Colletotrichum*. The conidia are unicellular, oblong and either straight or slightly curved. The fungus is carried to the land mainly by the spotted seeds, which contain dormant mycelium, it grows up with the growing plant, and extends to other plants by means of its spores.

**Red rot disease of Sugarcane:** Red rot of sugar cane is caused by *Glomeralla—tucumanensis* which is a perfet stage of pathogen. The red rot disease of sugarcane caused by *C. falcatum* **conidial stage** occurs throughout India, and also in other countries wherever the cane is grown as a standing crop. In Java, it is often referred to as red smut disease. It is one of the very serious diseases, that affect the sugarcane sometimes, resulting in complete loss of the entire crop.

The first symptoms appears on the third or the fourth leaf from the top of the plant which exhibits withering from the tip extending downwards throughout the margin leaving the center green. Gradually the entire foliage crown withers and droops down. In late stages appear all along the midrib of the leaves producing the characteristic symptoms. These contain the acervuli and conidia of the fungus. Ultimately the whole cane dries up, becomes light and can be easily broken. At this stage, minute, black velvety acervuli containing conidia are developed on either side of the nodes and sometimes along the depressions of the internodes also.

Internal symptom are seen by splitting open the canes. In early stages of infection, the vascular tissue of one or few of the basal internodes becomes red coloured emitting a sour smell. It than gradually extends over the pith also occurring as irregular longitudinal streaks or patches. Later the pith dries up completely and changes to dull brown colour. The irregular cavities developing in the pith are filled with mycelial network of the fungus. The discoloration is due to the cell walls turning red and the contents forming a coloured granular mass. The discoloured vessels may contain a reddish gum like substance. The stromatic hyphae under the rind rupture and at the nodes characteristic acervuli appear externally.

Infection of the host plants may be secured through:

(*i*) conidia and mycelium occurring in soil, air, or irrigation water from the diseases plants of the previous crop.

(*ii*) diseased sets employed for cultivation.

(*iii*) chlamydospores like appresoria formed copiously on the old mycelium within the diseases canes. These may go into the soil through the rotting of the cane.

The germ tube of the conidia, when growing on the host, swells up at the tip to form an appresorium at the point of contact with the host surface. An infection thread developing from the appresorium penetrates cuticle and develops further. The infection is established through wounds exposing the pith or scars created by tearing off the leaves or young buds and shoots are also through adventitious roots.

**Moniliales**

***Cercospora:*** Several leaf spot diseases of major economic importance are caused by species of *Cercosora. Cercospora apii* is destructive on celerly. *Cercospora beticola* on beet root. One of them *Cercospora nicotianae* causes "**frog eye of tobacco**". The disease gets its name from the appearance of the lesion, which when fully developed consists of a small, balanced area surround by a brown necrotic zone. Numerous lesions may develop on leaves in the field and seriously reduce their market value, but there is a second phase which is equally more important. Numerous brown spots appear on leaves when they are cured although they had comparatively few visible lesions at harvesting. Fungus survives on small pieces of leaf debris and unless precautions are taken, initiates the disease in seed beds.

Two species cause leaf spots on ground nut are *Cercospora arachidicola* and *C. personata*. The ascospores states have been found in the U.S.A. and described as *Mycosphaerella arachidicola* and *M. berkeleyii*.

*Cercospora* is a facultative parasite living in the soil and often becoming a destructive parasite causing commonly the leaf spots (Fig. 12.2 A-B). It is a large genus embracing nearly 3800 species. *C. apii* was the first to be described from the leaf spot termed as early blight disease of celery. Some species are also pathogenic on sugar beet, lettuce, tomato, potato, cucumbers, cherry and melons. In most cases the infection initiates as pale green spots on the upper surface of the leaf which gradually turn yellow ultimately, becoming brown. Often the fruits are not formed or remain small. The tikka disease of groundnuts caused by *C. personata* is one of the common and serious disease.

**Vegetative Structure:** Mycelium comprises septate and branched hyphae. In culture the young mycelium is colourless later on assumes greenish brown colour forming a velvety pad over which the conidia are formed copiously. In the host tissue, the parasitic hyphae are slender and branched, being intercellular in the leaf tissue with lobed haustoria penetrating the cells. At speculation time, the hyphae from small stromata under the epidermis, generally on both surfaces of the infected leaf. Short unbranched conidiophores emerge out by the rupture of the epidermis and bear single conidia at their tips (Fig. 12.2 C).

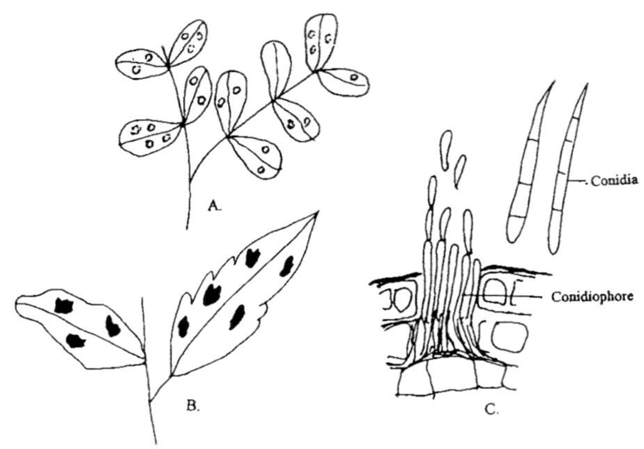


**Fig. 12.2:**
A. *Cercospora personata*. B. *C. arachidicola*. C. T.S. of leaf showing conidia on conidiophore.

**Reproduction:** *Cercospora* commonly reproduces asexually by means of conidia. In a few cases, however, the perfect stages have been found to belong to some genera in the Ascomycetes e.g. *C. cerasella* has been shown to be the conidial stage of *Mycosphaerella cerasella* and the perfect stage of *C. personata* is *M. berkeleyii*. Conidia are generally formed singly at the end of short vertical hyphal branches the conidiophores. They are small, unbranched, generally aspetate and some what thicker than the parent hypha. When the conidium has been cut off, the conidiophore forms a characteristic bend that shifts the conidium laterally on one side and finds its way to grow further to repeat the process. Sometimes 2–3 conidia may be seen attached in a row over such conidiophores. When ripe, the conidia fall of. Each conidium is inversely clavate *i.e.* rounded at the base and tapering apically. It may be straight or slightly curved and generally 4–5 septate. Conidia are ash gray to light brown in colour.

***Fusarium:*** Various species are specially important in seedling diseases of cereals and often more than one species is involved. The pathogenic activities these fungi are not always confined to seedlings. *Fusarium culmorum* also causes a foot rot and blight of the ears; so does *Gibberella zeae* is best known as the causal agent of wheat scab in which the heads and grains become infected. A progressive invasion of the plant from the seedling stage produces symptoms which have been given various disease names e.g. spring, yellow foot rot, leaf, ear and head blight. Infection of ears can lead to contamination or infection of seed, and within this group infection of young plants may arise either from this seed borne inoculum or from mycelium present in the soil.

Fusarial wilt fungi are all forms of *Fusarium oxysporum*, whose parasitism is limited usually to a single host species. F. sp. lycopersici on tomato, F. sp. *vasinfectum* on cotton, F. sp. *pisi* on peas, F. sp. *conglutinans* on Cabbage. F. sp. *Cubense* on banana. Infested soil is major source of inoculum and local dissemination of these fungi results from the movement of this soil in drainage water and on implements. Most of them remain in soil for a longtime in the form of chlamydospores.

These fungi invade the root cortex but do not damage it to any great extent except under special conditions. Then they become established in the vessels and are mostly confined there. At this stage there may be some spread in the field from an infected plant to an adjacent, healthy one especially where there is root contact (Fig. 12.3 A-C). The large amount of inoculum released into the soil places neighbouring plants at risk. In susceptible hosts the vascular system is often extensively colonized, even the leaf petioles and the production and distribution of bud conidia within vessels accounts for this. Symptoms appear as the water conducting system is invaded and cover the range epinasty, yellowing, vascular browning, tyloses, gums and wilting.

Various environmental factor influence the degree to which the fusaria colonize their hosts and therefore influence symptom expression. The fusarium wilts are favoured by high soil temperature. For tomatoes the optimum for *Fusarium* wilt is 28°C, disease development is poor below 21°C and above 33°C.

**Vegetative Structure:** Mycelium comprises a close network of profusely branched, hyaline hyphae that are septate and multinucleate, with age, the mucelium may exhibit various colours and may appear to form a felt over the culture media. The hyphae of the parasitic species, that penetrate the root tissue of the host to reach and ramify in the xylem vessels and tracheids are thin, septate and hyaline. These are branched and sometimes, plug almost the entire lumen of the vessel in which they occur (Fig. 12.3 A-C). In extreme cases of infection a white crust of mycelium may form over the external surface of the host also and the spores are produced over it.

**Asexual Reproduction:** *Fusarium* reproduces only asexually by producing chiefly, two kinds of spores.

(a) **Macroconidia:** Which are large, hyaline cresent-shaped (fusiform) spores that are pointed at both ends. These are septate with 3–5 tansverse septa and are formed over short, simple or branched conidiophores. They are generally, shed off when mature but sometimes, stick together as a slimy mass over the tuft of conidiophores in the sporodochium. In parasitic form the macroconidia are generally formed externally over the host surface (Fig. 12.4 A).

(b) **Microconidia:** Which are small-hyaline, elliptical ovoid or curved unicellular spores are usually held together in a drop of liquid over the hyphae. In parasitic forms, those are formed generally, with in the host tissue without collecting into masses. They resemble to the conidia of *Cephalosporium* and hence, often referred as the *Cephalosporium* stage of the fungus (Fig. 12.3 D).

In addition two spore forms chlamydospores (Fig. 12.3 E) are also formed over older hyphae, whether in culture or with in the host. The chlamydospores are ovoid to spherical structures formed by the rounding up of the cells from which they develop. They have thick walls and occur either, singly or in chains and their position may be terminal or intercalary. These spores are very durable and retain their viability for the long time.

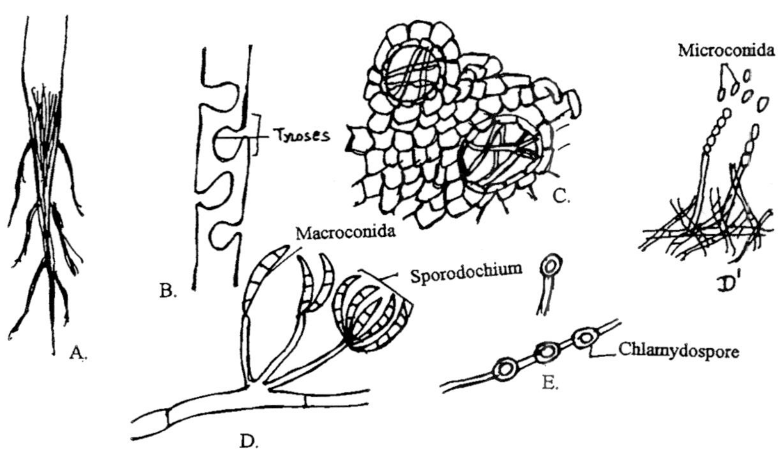


**Fig. 12.3:**

A. Root rot by *Fusarium*. B. Tyloses formation by *Fusarium*. C. Hyphae in the vessels. D. Sporodochium of *Fusarium*. E. Chlamydospores of *Fusarium*.

***Helminthosporium:*** This group is chiefly associated with seedling disease of cereals. *Helminthosporium avenae* (perfect state *Pyrenophora avenae*) causes a seedling blight of oats and also a leaf strip of older plants. The fungus is carried mainly as mycelium on the seed. *Helminthosporium graminearum* causes a similar seedling disease and leaf stripe of barley. It is also mainly seed borne. *Helminthosporium sativum* causes a seedling blight of barley and wheat. *Helminthosporium oryzae* (perfect stage *Cochliobolus miyabeanus*) causes a seedling blight and a leaf spot of rice and is also both seed borne and soil borne.

**Vegetative Structure:** The mycelium comprises hyphae that are branched, septate and coloured light or dark brown. In culture, the younger hyphae are slender, colourless and distantly septate but with age. They turn brown become wider and closely septate, some of the cell may become globular. In old culture or in dried up media, the hyphal cells may become thick walled, barrel shaped. They are brownish, highly resistant and may form even mycelial clumbs. With in the host, the parasitic hyphae are intracellular as well as intercellular without pronounced haustoria, penetrating the adjoining cells. The hyphae form a sort of stromata from which short, erect and septate conidiophores emerge out through the stomata or between the epidermal cells, either singly or in clusters.

**Asexual Reproduction:** As an imperfect form *Helminthosporium* reproduces commonly by means of conidia. There are many species whose perfect stage has yet not been found but a number of species of *Helminthosporium* are now being regarded as the conidial stages of some of the Ascomycetes. Those, whose perfect stages have been recognised are *H. iridis*, causing Iris leaf spot.

Conidia are formed generally singly at the tip of conidiophores which are upright, repeated, branched or unbranched, short, brown hyphal branches, arising from the parent mycelium lying sub epidermally (Fig. 12.4 B). They are generally formed in clustures and may occasionally be produced singly also. The conidia when mature, may either fall off and new one may be similarly formed over the tip of the conidiophore or quite often the conidium is pushed aside and the tip continues growth after a small bend at the point of attachment of the previous conidium. It may thus bear a few characteristic knee bends, each of which denotes the earlier position of conidial attachment. The conidia are subhyaline and slightly ovate when young but turn yellow brown or even dark brown at maturity. They are multiseptate, cylindrically elongate, widest at the base and tapering apically but with blunt ends. They readily germinate in water by forming germ tubes that may develop from any of its cell but generally they are polar.

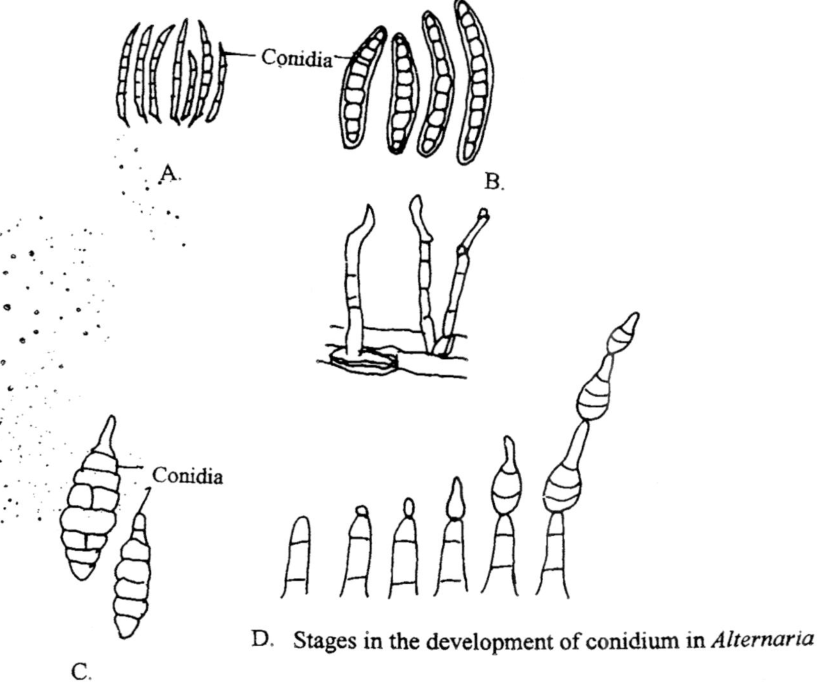


**Fig. 12.4:** Various types of conidia

A. *Fusarium* macroconidia. B. *Helminthosporium*. C. *Alternaria*. D. Stages of development *Alternaria* conidia.

Besides the conidia, in some forms e.g. *H. teres* the hyphae form small black sclerotia like structures that serve the purpose of perennation. Small spherical pycnidia containing unicelled, elliptical to round pycnospores are also known to develop on dried and starved tissues of barley infected with *H. teres*. Both of these on germination, form the new mycelium.

***Rhizoctonia:*** Many damping off diseases are caused by a fungus which on agar media produces a mycelium of hyphae 6–12μ diameter with a characteristic mode of branching. Branches arise at right angles and are constricted at the point of attachment to the parent hypha, there are septa within the parent hypha on either side of this and the base of the side branch is swollen and also septate. At first the mycelium is hyaline or light brown but later darkens and produces irregular, dark brown and often aggregated sclerotic. Isolates with these general characteristics are referred to the form genus *Rhizoctonia* and most to one species *R. solani*.

***Cladosporium:*** One species, *Cladosporium herbarum* is very common on semi moribund material and frequently causes blackening of the ears of cereals, where these contain little or no grain as a result of root infection. *Cladosporium cladosporioides* f. sp. *pisicola* causes a leaf spot and scab like pustules on the pods of peas.

Tomato leaf mould is more important. Chloritic patches develop on the upper surface of leaves and the corresponding areas on the under surfaces become covered with a greyish mass of conidiophores which later turns brown. Many conidia are produced in a warm and moist atmosphere and they readily gives rise to secondary infection. The spores are very resistant to drying and low temperature and are able to survive between crops in small bits of leaf debris and on bricks and woodwork. Temperature and humidity largely determines disease severity. Optimum conditions for the leaf mould are 22°C and relative humidity over 95%. At 10–15°C severe infection occurs under humid conditions but the disease develops more slowly. At humidities below 80% disease development is also related. Humid conditions at right appear particularly important possibly because light retards germ tube growth.

***Alternaria:*** Fungi belonging to this genus are commonly found on leaf spots and necrotic tissue. Their idenfication is not easy because of the variation in individual species. This is true to *Alternaria tenuis* group which occur on a wide range of senescent and moribund tissue. Other species are more pathogenic but again disease severity is associated with crop maturation. An example in *Alternaria solani* on potato and tomato. This disease is often called "early blight of potato" (Fig. 12.5 A). Another disease 'target spot' refers to the concentric rings in the leaf lesions and this occurs fairly common in *Alternaria* leaf spots. The leaves of severely infected potato and tomato plants drop prematurely and yields are decreased. *Alternaria solani* lives for at least a year on old vines and may also persist on other perennial solanaceous hosts. Spores from these sources infect tomato and potato either by direct penetration of the article though stomata.

*Alternaria brassicae* and *A. brassicicola* are other two species frequently encountered. They are associated with dark leaf spot of brassicas and can sometimes be troublesome on cabbage and cauliflower plants grown for seed. Lesions on the infloresences cause the pods to rupture prematurely and much seed is lost. These fungi are seed borne and some mycelium is deep seated. They can be eliminated by immersing seed for 18 minutes at 50°C.

**Vegetative Structure:** The mycelium comprises hyphae that are profusely branched and septate. In culture they are hyaline to start with becoming olive buff, or olive green or light brown with age. The branches in culture are long white, in the host they may be short. The hyphae in parasitic species, are both inter as well as intra cellular and are generally brown (Fig. 12. 5A).

**Reproduction:** *Alternaria* reproduces by means of polymorphous conidia. These arise as small buds over tips of the conidiophores (Fig. 12.4 D). They are much variable in shape and structure but in general, the conidia are long, bottle shaped with a broad dark coloured cup like base and a long narrow light coloured neck commonly called at its beak. The body of the conidium may be divided into small compartments by 5–10 transverse septa. A few of the larger partitions of these may be further subdivided by a longitudinal septum. The younger conidia may be hyaline, ovate and single celled (Fig. 12.5 B).

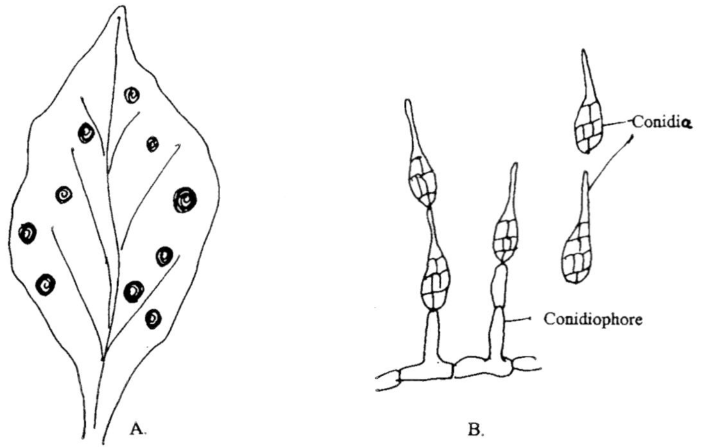


**Fig. 12.5:**

A. Habit sketch of *Alternaria* on Brassica leaf. B. Conidia present on conidiophore.

The conidiophores in culture, are generally long over which a chain of such conidia, may be formed but in parasitic forms, the conidiophores are usually short emerging form the stomata or generally from the older conidiophores, conidia are never formed in long chains being commonly borne singly or sometimes in chains of two conidia only. The conidiophores do not appear over the freshly invaded portions.

Conidia when mature, get detached and are disseminated by wind. Falling on suitable substrata, they germinate by germ tubes developing from one or more of these cells of the conidium. In parasitic forms, the germ tubes penetrate through the stomata or weak and injured host epidermis developing visible leaf spot symptoms with in 2–3 days of the infection.

# 13. LICHENS

The lichens are a group of plants having a composite structure consisting of two widely dissimilar organisms–a fungus and an algae-associated in symbiotic union in which, the fungus is the more predominant partner furnishing the reproductive part.

**History:** Theophrastus (371-284 B.C.) disciple of Plato and Aristotle in his book 'History of Plants' was the first to use the word 'Lichen' for extra plant growths on the tree barks. Until the early 17th century the lichens also designated to bryophytic plant forms like *Marchantia* and the mosses. Morison, in 1699, called such plants as 'Musco fungus' while Tournefort (1700) described lichens as plants with a shallow cup like fruit but lacking flowers. A few lichens were placed under the genus Coralloides but Dillenius (1741) preferred to use the term Lichenoides for all the lichen like plants. Linnaeus in his 'binomial nomenclature' considered lichens as an independent group of the Cryptogams and tried to group the genera on their thallus characteristics.

From the early period of the 19th century work started gaining significance towards the study of the morphology, anatomy and physiology of the lichen as a whole. Schwendener in 1867, laid a firm foundation for a study of modern lichenology by enumerating dual hypothese's which postulates the composite nature of lichen thallus to be composed of forms resembling Algae and Fungi. He defined lichens to be merely fungi parasitising algae. This view was opposed by the lichenologists of that time namely Lauder Lindsay (1959), Crombie (1885) *etc.* Schwendener's dual hypothesis was finally discarded when Rees, for the first time in 1871 obtained success in producing a synthetic culture of Lichen thallus by showing spores from the apothecium of genus *Collema* with pure cultures of *Nostoc*. He also found that the hyphal fragments disorganised if *Nostoc* chains were not available. Stahl

(1877) added further evidence to show that both these organisms were of independent origin when he developed the Lichen thallus complete with fruit bodies and spores by synthesising culture of mature spores with algal units present in the hymenium.

Reinke (1872), described the connection between the fungus and algae as consortium *i.e.* an association in which both have mutual growth and interdependence. deBary (1879) took up this idea of mutual advantage still further and proposed the word symboisis or conjoint like benef- 'al to each of the partners.

**Habit and Occurrence:** The lichens are cosmopolitan occuring on a very wide variety of habitats ranging from the seashore to the high mountain tops. They generally do not thrive near the large cities where the atmosphere is polluted by the industrial or residential smoke. The lichens are mostly perennial, aerial, slow growing and long lived plants. They usually abound in those places also which appear unsuitable for the normal plant growth *viz.* bare hard rocks and the cold Arctic regions. *Cladonia rangifera*, commonly known as the 'reindeer moss' grows luxuriantly in the tundras. The most frequent abode of the great many lichens is on the leaves, bark of the tree trunks or even on the soil, where there is abundant moisture available as in the humid tropical forests. The lichens are of various colours–white, yellow, brown, grey or even black.

In the symbiotic association, the algae perform the photosynthetic function, because they have chlorophyll and do the job of assimilation by manufacturing carbohydrates. While the fungus partner thus, gets a supply of sugar or carbohydrates for its nourishment from its algal consort, it in turn provides, protection, water, nitrogenous substances and salts absorbed by it. The rhizoids when present, mainly perform the absorptive function and are fungal in origin. In addition, they also serve to anchor the thallus firmly to the substratum.

**Classification:** The lichens have been variously classified and the three important criteria on which the general classification is based are given below:-

1. On the basis of the fungal component, Lichens are divided into two subclasses.

(a) **Ascolichens:-** In this association fungus belongs to the Ascomycetes. The reproductive structure and fructifications are ascomycetous in origin and structure. The Ascolichens are further classified into 2 series depending upon the structure of the fruit body. *viz.*,

(i) Gymnocarpae: includes all such lichens whose frutification is an apothecium *i.e.* in the form of a more or less open disc *e.g. Parmelia.*

(ii) Pyrenocarpeae: includes those lichens where the frutification is a perithecium *i.e.* a closed structure *e.g. Dermatocarpon.*

The algal components of the Ascolichens belongs to two main divisions of Algae *viz.* Myxophyceae, and Chlorophyceae. Cynophyceae occur mainly in the gelatinous lichens, while the Chlorophyceae in leathery forms. The algal components in their association to form the lichen thalli, are generally not affected much except in a few cases which may either lose their colour *Gloecapsa* or that the cells of the trichomes may become loose (*Nostoc, Scytonema*).

(b) **Basidiolichens:** The fungus belongs to the Basidiomycetes. The reproductive structures and frutifications are basidiomycetous belonging chiefly to the family Thelephoraceae in Autobasidiomycetes. There are only 3 genera belonging to this group of lichens which occur generally in tropical countries, mainly in South America. These genera are *Cora, Corella* and *Dictyonema*. The first one being more common. The algal element in the Basidio-lichens generally belongs to the Myxophyceae group corresponding to either *Chroococcus* or *Scytonema*.

2. On the basis of algal components in the thallus Lichen thalli are distinguished into 2 categories.

(a) **Homiomerous**: Thalli in which the algal cells and fungal hyphae are uniformly dispersed through out the thallus. In such forms, algal constitutent does not dominates and is usually gelationous belonging to the Myxophyceae. Fungus element grows side by side (Ephebe) but sometimes, a sort of thin protective layer may be formed outside the thallus or covering the apothecium *e.g. Collema.*

(b) **Heteromerous:** Thalli, in which the algal cells are restricted or confined to form a distinct layer generally on the upper side of the thallus. In such forms fungus constituent leads in the thallus formation. The thallus structure is differentiated into distinct layers of the cortex, algal zone and the medulla, the first and last being composed of hyphae *e.g. Parmelia* the algal constituents here belong chiefly to the non gelatinous Chlorophyceae.

3. On the basis of the type of thallus and their mode of occurrence, lichens are generally classified into 3 categories.

(a) **Crustose:** 'encrusting Lichens' which occur as their or thick crusts over rocks, soil or tree bark. The crust may sometimes, be partially o completely embedded within the substratum but generally, it occurs as a superficial layer. *Rhizocarpon* grows on rocks and *Lecanora, Graphis etc.* on barks (Fig. 13.1 A).

(b) **Foliose:** Leafy Lichens which occurs as horizontally spreading leaf like lobed structures which attach firmly on the substratum by means of special rhizoid like organs developing from the darker lower surface of the thalli. There may either be a central point of attachment as in *Umbillicaria* or generally several where the thalli are much folded and lobed as in *Parmelia etc.* Rhizoids are fungal in origin. They spread both by marginal and apical growth (Fig. 13.1 B).

(c) **Fruticose:** 'Shrubby lichens' which occur as a cylindrical, flat or ribbon like upright, generally branched strcutures attached to the substratum by their basal ends. The thallus of a fruticose lichen varies from short, tough, flat, lobe like strcuture in some species of *Ramalina* to long. Cylindrical, hanging down, thread like thalli of *Usnea, Alectoria etc. Cladonia* is an intermediate genus to these two forms of thallus in which there is an evanescent, crustaceous or lobed primary basal thallus which arises as an upright, vertical, simple or branched, secondary thallus called podetium. This sometimes forms a cup like 'scyphus' apically over which the apothecia are borne (Fig. 13.1 C).

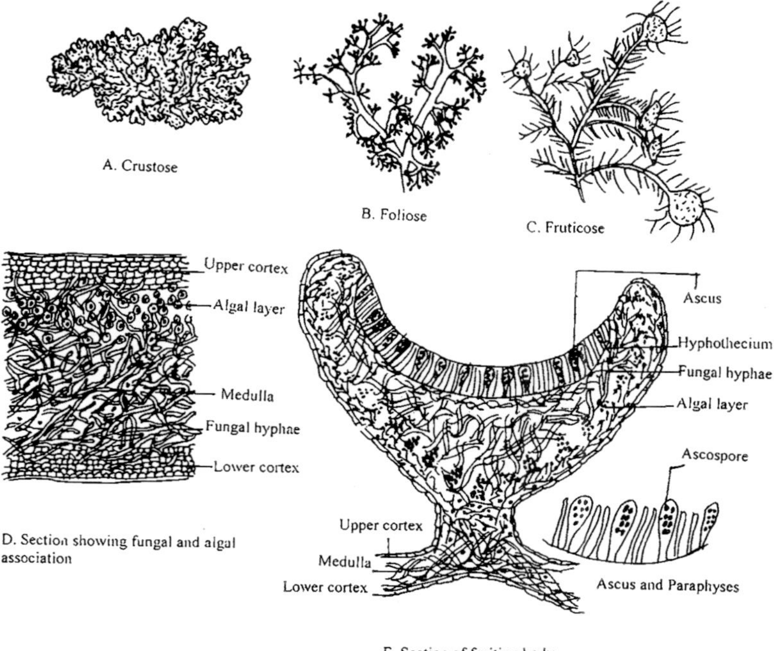


**Fig. 13.1 :**Various types of Lichen (A-C)
D. T.S. of thallus showing internal structure. E. Apothecium

**Vegetative Structure:** The two symbionts *viz* algae and the fungi are so associated together as to form a simple thallus of undifferentiated tissues in its structure. Many of the algae are unicellular and variously distributed with in the lichen thallus while the real framework of the lichen body is made up by the branching hyphae of the fungus component which become modified variously to perform the protective, strengthening or reproductive functions. Morphologically, the Lichen thalli may either be crustose, foliose or fructicose types. The typical structure of most of the Lichens is revealed in a section of the quite commonly occurring foliose lichens in which the following regions constituting the thallus can be easily recognised (Fig. 13.1 C).

(i) **The Upper Cortex:** Formed by compactly inter-woven fungal hyphae enclosing either almost no interspaces between them or if pesent the interspaces are filled with gelatinous substances. There is usually no epidermis like layer present externally but quite frequently the upper cortex is covered by a layer of homogenous mucilage forming the cuticle. Sometimes, a felt of thick walled, septate, anastomosing, branched or unbranched hair may be present on the upper surface *i.e. Peltigera, Parmelia etc.*

In some foliose Lichens (*Parmelia, Ramalina*) develop breathing pores which connect the inside of the thallus with the outside atmosphere. These help and facilitate the gaseous exchange in the thallus. These may either be conical or in level with the thallus on its upper surface and are continuous with the upper cortex but the hyphae are loose in the region of a 'breathing pore'.

(ii) The Algal layer, occurring just below the upper cortex forms the photosynthetic zone of the thallus. It remains either as a continuous layer or broken into scattered patches of green myxophycean or chlorophycean algal elements. In *Solorina, Crocea* the algal cells occur as pyramids rising and pointing towards the upper surface. This layer is also called as gonidial layer because of the earlier concept to regard these cells as gonidia *i.e.* having a reproductive function though in reality they have no such function. The fungal hyphae in this zone have thin walls and occur loosely between the algal cells.

(iii) The Medulla, occurs nearly in the middle of the thallus beneath the algal layer forming a mjor portion of the thallus. The hyphae have smooth and thin walls and are loosely interwoven.

(iv) The lower cortex, like the upper cortex comprises of closely packed dark colored hyphae either running perpendicular or parallel to the lower surface. The rhizoids arise from this layer.

In many of the crustose thalli as in *Lobaria pulmonaria* the medulla is uncovered by the lower cortex. It may be exposed to the atmosphere or may rest directly on the substrate few crustose thalli exhibit a thin undifferentiated structureless mixture of algal and fungal elements.

**Special Structure:** Four types of special structures are frequently found on lichen thalli.

(i) **Soredia:** It occur on the upper surface or margin of the thallus as powdery dust or with in definite pustule like compact strcutures called soralium. The soredia are separable portions of the thallus consisting of one or more algal cells clasped/entangled or surrounded by fungal hyphae both belonging to the same thallus. A soredium hyphae develops on a hyphal branch from the algal layer which grows out and envelopes one or more newly divided algal cells. The elongating branch pushes its way out and finally gets detached from its support. The soredia are thus easily disseminated by wind or other agency and falling on suitable substrate, develop into a lichen thallus (Fig. 13.2 C).

The soralia arise from the algal layer below the upper cortex. The algal cells divide actively to be soon surrounded by the thin walled hyphae present interlacing these cells. Their growth in the upward and outward direction breaks open the cortex at outward direction which breaks open the cortex at definite points. The position and size of the soralium varies greatly even with in a single genus *e.g.* in *Parmelia* they may either be as small round dots on the surface or as irregular elongate furrows or as pearl like or tubercled emergences towards the margin.

(ii) **Isidia:** Also occur on the upper surface of the lichen thalli as minute or coral like, simple or branched out growths. The algal element within the isidia is the same as that of the parent thallus and is covered by a definite cortex which is continuous with that of the thallus. The function attributed to the isidia is to increase photosynthetic surface of the thallus. These also sometimes act as organs of vegetative propagation, if detached from the parent thallus *e.g. Peltigera* (Fig. 13.2 C).

(iii) **Cephalodia:** Occur as a gall like out growths generally on the outer surface of the thallus except in a few cases where they are endogenous *i.e.* endotrophic as in *Nephroma*. The cephaloida are usually dark coloured occurring either, as a flat orbicular discs or as coralloid banches or as irregular warts and tubercles. The cephalodia are covered externally by the cortex of the parent fungal hyphae but differ from isidia in having the algal constituent different from that of the parent thallus which generally belongs to the myxophyceae group. The cephalodia are of help in retaining moisture (Fig. 13.2 A).

(iv) **Cyphellae:** Occur on the lower surface of the thallus, quite common in the genus *Sticta* as small, hollow, circular, white depressions with its base resting on the medulla and its margin formed from the rupture cortex projecting slightly inwards. Their chief function is to facilitate exchange of gases between the interior of the thallus and the exterior atmosphere (Fig 13.2

The cyphellae should be distinguished from pseudocyphellae which also occur on the lower surface of the thallus as roundish openings in the cortex through which the hyphae protude out. They are simple in structure, have no definite margin and called Pseudocyphellae.

**Asexual Repoduction:** Asexual reproduction in the lichen takes place by the following methods:-

(1) **Fragmentation:** Breaking of the thallus into small fragments, each of which regenrates to form a new thallus.

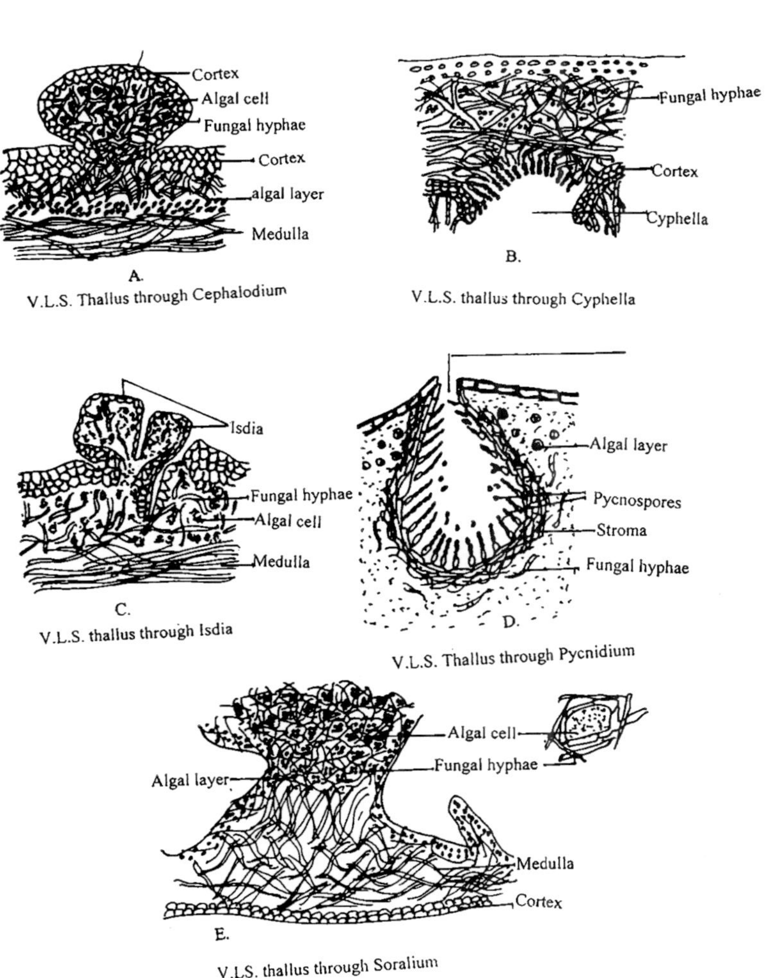


**Fig. 13.2:** Various types of structure found in lichen thalli

(2) **Soredia and isidia**

(3) **Rejuvenation:** Growing young again. This is a common method in species of *Cladonia* where the basal portion of the podetia die off and the branching continues above.

(4) in addition to the above methods–lichens may reproduce asexually by conidia, oidia and pycnospores. These are fungal in origin and germinate into hyphal structures which in contact with suitable algal consort, develop into a new lichen thallus.

**Sexual Reproduction:** Lichen thalli are composed of two different organisms *i.e.* an algae and a fungus of these it is only the fungus element that subscribes to the sexual reproduction for the thallus. Accordingly the sexual reproduction in all the Ascolichens corrsespond in principle to the ascomycetous sex organs in origin and structure while in the Basidiolichens they are basidiomycetous.

Fuisting (1868) was the first to demonstrate the existence of ascogonial structures which were connected with the development of the fruit bodies. Stahl (1877) Baur (1898) and several others showed the existence of female carpogonia and male spermogonia. The later are flask shaped, pycnidia like structures on the thallus generally scattered all over or restricted to the periphery of lobes (*Parmelia*) or at the margins (*Cetraria*) or the tips of Podetia *e.g. Cladonia.* The spermogonia originate from hyphae in or near the algal zone and remain immersed in the thallus. They may assume, ovate, pyriform or globose form. The inside of the chamber may either by smooth and round or thrown into inwardly projecting folds. The interior is lined by simple, branched or unbranched aseptate or septate hyphae the spermatiophores, which bud off spermatia. The margins of the spermogonia may be lined by sterile paraphyses which emerge out through the apical ostiole above the thallus. The maturre spermogonium is filled with numerous spermatia and with abundant mucilage which swells with moisture and helps in their explosion out.

The spermatia are minute of varied form and size but more or less cylindrical, often ellipsoidal, oblong, dumbbell shaped, accicular, straight or bent. They are usually colorless, surrounded by a cell wall and have a distinct nucleus.

The carpogonia resemble in basic structure with those of the ascomycetes *i.e.* having a multicellular elongate trichogyne followed by a helically coiled ascogonium. In majority of the genera, the ascogonium lies embedded in the thallus at varying depths near the cortical region. It is many celled, the cells may be in spiral or in irregular groups and generally uninucleate but sometimes mutlinucleate as in *Peltigera* and *Solorina*. The slender trichogyne elongates vertically upwards to emerge its tip out of the surface of the thallus. The tip is usually sticky being surrounded by a mucilage sheath as in *Physcia* (Fig. 13.2 D). The spermatia on dissemination fall on the tip of the trichogyne and its nucleus migrates, throughout its length to reach the ascogonial cell. After fertilistation the trichogyne cells loose turgidity and their septa swell up considerably to appear as knots in the hypha. The ascogenous hypha develop from the fertilized cells of the ascogonium which form the asci and ascospores in the same manner as in any of the ascomycetous fungus.

The carpogonia may occur singly and produce the ascocarp (*Collema*) or they may occur in numbers but only one out of the lot is fertilised (*Usnea*) while, in others several of them share in the formation of the apothecium (*Lecanora subfusca*).

In *Collemodes bachmannianum*, the spermatia are not produced in distinct spermogonia neither the trichogyne projects vertically upwards beyond the thallus. Instead it elongates horizontally with in the thallus until it reaches the spermatia which are produced in clustures and are budded out laterally form hyphae almost in similar positions where the spermogonia would have been formed.

An empty spermatium attached to the tip cell of the trichogyne in either of the cases, indicates the transfer of male nuclei into the ascogonium through the trichogyne, its actual transfer and migration have not been demonstrated.

In every case, the ascogenous hyphae start developing from the fertilised ascogonium cells although in some cases the development occurs parthenogenetically, *i.e.* in *Phlyctis agelaea* and *Peltigera* the spermogonia are either lacking or are rare and in the latter genus the trichogyne are normally not present. Such cases also occur in species of *Parmelia, Gyrophora, Solorina etc.*

**Development of Ascocarp:** The ascogenous hyphae are profusely branched with cells having either one, two or many nuclei. In most cases, the penultimate cell or the cells towards the apex are dikaryotic with the typical crozier formation. This is the young ascus in which the two nuclei fuse and with three successive divisions eight ascospores are formed with in the mature ascus. The first division is reductional, there is simultaneous development of a large number of sterile hyphal branches forming the paraphyses in between the asci. Alongwith a sterile protective hyphal cover from the parent thallus, grows to envelope the ascocarp either without or with the algae layer intact. The ascocarps may either be perithecium type, when they are immersed in the thallus and open out by an apical ostiole when they are flattened, discoid, cup like structures over the thallus, typical of the Discomycetes (Fig. 13.1 E).

The apothecial development may be of two types (1) those which are closed at first probably due to the very tough layer *e.g.* in *Lobaria*. In such cases, the early stages of development resemble the perithecial structures and (2) those which remain open from the very beginning.

The mature apothecia may either be covered only by the fungal hyphae which form the proper margin or, by the thallus including algal constituents also forming the thalline margin. The internal structure of mature apothecium comprises of 3 distinct zones *i.e.* the thecium is the fertile zone and is composed of the asci interspersed with paraphyses which generally project beyond the level of the asci to form a layer called epithecium and the hypothecium which is composed of loosely packed hyphae below the thecium. The hypothecium layer may be light coloured or dark coloured.

The asci usually contain 8 ascospores with in but sometimes, the number of spores per ascus may be few *e.g.* 2 in *Endocarpon*, 1 in *Lopadium. Umbilicaria* and sometimes many as in *Acarospora, Biatorella*. The ascospores of lichens are of varied size, colour and forms, single celled to one septate or multiseptate, generally uninucleate, but sometimes multinucleate as in *Lecidea pertusaria* etc, which on germination, produce a number of germ tubes all over the spore surface. The outer wall of the ascospores may be either smooth or may be variously thickened, sculptured, pitted or may show reticulations, markings and warts. In a few lichens, the spores are polari-bilouclar *i.e.* 2 celled with the medium septum thickened so much that the lumen of each cell is reduced to small area at the polar ends. All the asci and ascopores do not mature dehisce or disseminate simultaneously but the process continues for long, may be for several years depending upon the conditions. The spore discharge from the mature ascus is effected by two fold pressure of the paraphyses and the marginal hyphae on the receipt of moisture. They may be shot upto 1 cm. distance. On germination the spores form germ tubes, generally one form each cell of the spore which, on coming in contact with the suitable algal element, form the new lichen thallus.

**Succession**

Lichens are well known as pioneers in plant succession. They virtually occur in every pioneer terrestrial habitat from polar regions to tropical areas and in many deserts where they are able to form long lived stable communities, sometimes locally exceeding the biomass of higher plants. Lichen succession in largely directional, and changes in the environment determine the ultimate fate of the communities. Stages in the successsion can be arrested only if the environment either remains unchanged or is not subject to change. These conditions are most easily met in the polar regions deserts, roct out crops in temperate areas where growth of trees is prevented, and rocks along rivers, lakes or oceans. In these habitats lichen communities can last for centuries. However on substrates such as in habitats invaded by shrub or tree canopies, the pioneer communities are eventually replaced by mesophytic lichens.

Succession among corticolous communities mainly depends on the photophyte (host tree of an epiphyte). The *Lobarion* & European climax community which seems to be associated with climax woodland consists largely of foliose lichens. There is a considerable variation among the species in succession of soil. In west Greenland it was found that sitt first colonized by about ten species of vascular plants before the first macrolichens appeared. The latter had reached about 2 cm diameter before the first crustose species could be identified. On the other hand *Lecanora epibryon* a crustose lichen is reported to figure prominently in communities on calcareous beach deposits in Spitzbergen. When a new rock surface is exposed there is also a recognizable series of successional stages. Most of the studies reveal crustose lichens as pioneers, followed by foliose and fruticose forms in an orderly succession. There is sometimes non-directional succession where rapidly growing foliose lichens are lost and replaced by crustose lichens. This leads to a kind of recycling of the communities.

There role of lichens in succession illustrates many of the principles of ecosystem development. Although compared to many organisms they contribute little to energy flow in a community, they are still significant in photosynthesizing in many harsh ecosystems such as deserts and rocky temperate mountains where they are apt to increase the utilization of incident light energy. Species diversity is often considered to reach its peak just before the climax and then a decline slightly. However in densely populated areas such as Western Europe, the presence of a rich lichen flora often indicates relict areas of primeval forest as in the New Forest, England. As succession proceeds, diversity of lichens accompanying fungal epiphytes and parasymbionts, provides a suitable habitat for a considerable microfauna and enriches the microflora as well.

**Ecophysiology of Lichens**

The lichens grow in a wide variety of habitats. They are commonly found growing on the old walls, roofs of houses, leaves, tree bark, base earth and even barren, unpromising rocky surface etc. Generally they are xerophytic in nature and can withstand long periods of drought. Consequently they thrive and multiply in habitats where other vegetation does not exist such as sand dunes. Deserts and rocks. They absorb, whatever substances they require through the whole surface of the thallus.

Lichen phycobionts can be found in a free living state, but most of them are ecologically very much successfully in symbiosis with lichen mycosymbionts than when apart. Water and dissolved salts along with extra cellular fungal metabolites are transpired in the passively moving water flow along the fungal cell wall, on in this way reach the phycobionts cells. Mycobiont derived urea and phenolic secondary metabolites play significant role in cyclic regulatory process *Lecanorales*.

Lichens are most sensitive to temperature, moisture, light and nutrients of the habitat. Since they are growing in Xeric and stress conditions their ability to $N_2$ fixation is an important phenomenon.

**Temperature:** The temperature profile below the soil is of no significance to lichens. But surface temperature maxima and minima and the actual temperature gradient about the surface is largely utilised by crustose as well as many foliose and fruticose lichens.

**Moisture:** Lichens are poikilohydric and depend entirely on periods of rain or dew or high atmospheric humidity to achieve a satisfactory level of thallus hydration and a resultant vigorous rate of metabolic activity. They can be regarded as oppurtunistic, actively metabolising during periods of thallus hydration an with very low levels of activity during air. There is a close geographical correlation between the abundance of lichens and the distribution of rainfall. Habitats with strong continuous daily levels of solar radiation and hence high rates of evaporation resulting in a lichen growth. The absorption of liquid water is rapid by dry lichen thalli.

**Nutrition:** The lichen flora of limestone and that acidic igneous rocks are quite distinctly different showing importance of mineral requirements of different species. The floristic richness of corticolous lichens found on trees with nutrient rich bark and those which occur on back of low nutrient status again point to the importance of ionic environment of the lichen. Lichens are very sensitive to pollutants pH of the substructure is very important for lichen colonisation. Lichens with heterocystous blue green algal symbionts show amino acid and protein synthesis by nitrogenase activity.

**Economic importance of Lichens:** Lichens produce about 600 chemicals which helps them survive in marginal environments and formed by attack of bacteria, fungi and grazing herbivores. The lichen substances include pigments. Toxins and antibiotics are very useful to people in diverse cultures, especially as a source of dyes and medicines.

Many species of lichens are valuable source of food which have little nutritive value. The edible lichens are harvested and dried for human consumption of as fodder for cattles. The traditional use of lichens in the preparation of dyes deserves a speical consideration. The fungal components of certain species of lichens produce coloured pigments which have been used as dyes in colouring fabrics and paints. Litmus is also obtained from lichen which is widely used dye in chemical lab as an acid base indicator. Some species of lichens used as medicine to care many diseases and sweet scented thalli of some lichens are used in the manufacture of purfumeries.

In eastern himalayan region, very few lichens are in use by local inhabitants. The Santhal's in Northern Bengal and Assam and Jaintias in Meghalaya use many species such as *Heterodermia, Evernlastrum, Parmotrema reticulatum* and *P. tinctorum* etc. as species and flavouring sgents to increase the taste and fragrance of non vegetarian preparations, pulses and other vegetables. Lichens are also used in industry for aromatic resinoids. The bulk of lichens for this purpose, Commercially known as 'charilla', 'Salay phool' and 'Hara phool' are transported from Sikkim Himalayas and recently from Kameng district in Arunachal Pradesh.

# 14. MICROBES IN BIOTECHNOLOGY

Biotechnology can be defined as an industrial use of living organisms or their components or the technological advancement to know what is the potential of biological system or their products for the benefit of mankind. It is highly multidisciplinary; dependent not only upon scientists in biology and chemistry and engineers but also upon financial, legal and managerial experts. A lot of work has been done during the past two decades has revolutionised this exciting area of study. There is not a even single aspect of human life that has not been benefitted by it. The biotechnology has emerged as a very promising interdisciplinary field with abundant potential to overcome problems related to agriculture, health, industry, environment, energy and many other important areas.

The use of microorganisms for large scale industrial purposes is not new, although it has been given a lot of importance in recent years. Centuries ago, people in Asia and Africa learned to make wine, beer, vinegar and 'saki' with bacteria and yeast without knowing the scientific basis of such production. The technology related to microbial production of metabolites, such as ethanol, lactic acid, butanol, riboflavin, and enzymes, such as protease, amylase and invertase, was also developed in the early decades of the 20th century. Large scale production of the antibiotic penicillin, was perfected during world war II and the production of many other antibiotics, amino acids, nucleotides and enzymes had been successfully accomplished in the 1950s and later. In recent years, microorganisms have found their application not only in the production of a variety of metabolites but also in the biotrasformation of several chemicals. The genetically engineered microorganisms are also being used for the commercial production of some non-microbial products such as insulin, interferon, human growth hormones and viral vaccines. Microbes have been widely and successfully used for a very long time in industry and they play a crucial role in everyday products like cheese, yoghurt, pickles, beer, wine, etc. (Dimmling and Nesemann, 1985). These can reduce environmental pollution through a variety of processes and other means including: (i) recovery of metals from polluted water ways; (ii) elimination of sulphur from metal ores; and (iii) use of biofertilizers and bio-pesticides. In the energy sector, these can be used for production of single cell protein to solve food and fodder problems, and for biogas production to provide energy for electrifying villages. Through the use of biofertilizers and bioinsecticides, microbes can also add to crop productivity. Similarly, these can provide cheap and cost effective methods of mining and metallurgy. Thus, microbial biotechnology will have a great impact on industry in the 21st century.

**Production of Organic Compounds by Microbial Fermentation**

A variety of organic compounds, used as organic feedstock for many chemical industries, are produced industrially by fermentation process. Microbial production of organic compounds that can be used as substitutes for traditional fuels also received major attention in 1970s when petroleum and natural gas became scarce. For both feed and fuel, generally the carbohydrate rich plant products are used as substrates. It is estimated that about 95% of total of 180 + 109 tonnes of plant material produced annually on this earth remains unuitilized as human or animal feed. By evolving suitable technology, such a large amount of biomass can be converted into a variety of useful products (Mayer, 1986).

**Ethanol Production**

Ethanol for human consumption has been manufactured as a component of alcoholic fermentation since prehistoric times. In recent years it has also been used as an important chemical feedstock and as a fuel supplement in Brazil in the 1980s, 20% of petroleum imports have been replaced by ehtanol produced from sugarcane. In India, in two years, approximately 910 million litres of ethanol was produced against a consumption of 826 million litres. Ethanol is produced by fermentation of some sugar rich products with the help of yeast, *Saccharomyces cerevisiae* or some times with *Kluyveromyces fragilis.* Several other organisms are also known to produce small quantities of ethanol (Table 14.1). In alcoholic fermentation in addition to the main products ethanol and carbondioxide, number of other by products are produced. The most important of the industrial chemicals include methanol, higher alcohols, such as propyl, butyl and amyl alcohols, glycerol acetaldehyde, acetic acid and various other acids (Crueger and Crueger, 1982; Evèleigh, 1981; Haseguwa, 1985).

**Table 14.1:** Microbial species used for producing commercial products

| | |
|---|---|
| **Industrial Chemical** | |
| *Saccharomyces cerevisiae* | Ethanol |
| *Kuluveromyces fragilis* | Ethanol |
| *Clostridium acetobutylicum* | Acetone and butanol |
| *Aspergillus niger* | Citric acid |
| **Enzymes** | |
| *Aspergillus oryzae* | Amylases |
| *A. niger* | Glucamylase |
| *Trichoderma riesii* | Cellulase |
| *Saccharomyces cerevisiae* | Invertase |
| *S. lipolytica* | Lipase |
| *Aspergillus* | Pectinases and Proteases |
| *Bacillus* | Proteases |
| *Mucor pussillus* | Microbial rennet |
| **Pharmaceuticals** | |
| *Penicillium chrysogenum* | Penicillins |
| *Cephalosporium acremonium* | Cephalosporins |
| *Streptomyces species* | Amphotericin B |
| *Bacillus brevis* | Gramicidin S |
| *B. licheniformis* | Bacitracin |
| *B. polymyxa* | Polymyxin B. |
| *Escherichia coli* (Via recombinant DNA Technology) | Insulin, human growth hormone, Somatostatin. Interferon |

### Gluconic Acid

Gluconic acid and its salts are widely used in pharmaceutical, food, feed, detergent, textile, leather and photographic industries. During glucose oxidation, gluconic acid is produced by *Aspergillus* spp. (Paper and Fehnell, 1965) and several other species of *Penicillium, Mucor, Fusarium* and *Aureobasidium. Aspergillus niger* is the most widely used microorganism. It contains very active form of the enzyme glucose oxidase. *Aspergillus niger* produces citric acid. *Acetobacter* sp. produces acetic acid. *Saccharomyces cerevisiae* produces citric acid, while *Lactobacillus prueckii* and *L. balgaricus* proauce lactic acid. Vitamins are also produced by many fungi and bacteria. *Ashyba gossypii* helps in the production of Riboflavin, while *Pseudomonas denitrificans* produces Vitamin B12 (Herrick and May, 1928).

### Enzymes

Enzymes have found a variety of applications in medicine, food textile and leather industries. In U.S.A. first enzyme named as "fungal tikka disease" was produced. Now a days several enzymes are produced by various microorganisms. Alpha amylases constitute the most important group of enzymes which hydrolyse starch. These are produced and excreted out into the medium by many bacteria and fungi, including species of *Aspergillus, Candida, Mucor, Neurospora, Penicillium, Bacillus* and *Lactobacillus.* Lactase, converting lactose to glucose and galactose, can be more easily utilized by microorganisms. Cellulases have also found some use in food processing, but are now receiving intensive study in connection with the utilization of lignocellulose, the cheapest and best organic carbon sources. Most of these enzymes are produced by *Aspergillus niger* and *A. oryzae.* Cellulase is produced by *Trichoderma* spp. and *Chaetomium* spp. (Goodman, 1950).

Proteolytic enzymes vary with respect to their pH optima. In general, alkaline proteases are produced by bacteria and acid proteases by fungi. There is a large demand for proteases as detergent additives. An important acid protease is rennin, used in cheese manufacture to coagulate the milk proteins. Rennin produced by species of *Mucor, Aspergillus* and *Endotha* accounts for about one third of the market. Lipases are produced and secreted by many fungi and bacteria such as *Penicillium chrysogenum*, *Pseudomonas fragi*, *Rhizopus delawas*. *Aspergillus niger*, *Mucor* spp. and *Staphylococcus aureus*. Lipases hydrolyse fats into diglycerides, monoglycerides, fatty acids and glycerol.

**Alcoholic Beverages**

Under aerobic conditions, yeasts metabolise sugar to carbon dioxide and water. If oxygen is scarce or absent, or if the sugar concenteration is high, fermentation occurs with the production of ethanol and carbon dioxide. Alcoholic fermentation is the basis of the production of the great variety and huge quantities of alcoholic beverages consumed by man. *Saccharomyces cerevisiae* is the major yeast responsible for the vast majority of fermentation reaction's. Alcoholic beverages can, (on the basis of the procedures involved in their production,) be grouped into three classes. First are those, like wine and cider, which are made by fermenting plant juices rich in sugars. Secondly, there are those such as beer, made from plant materials rich in starch, which must be converted to sugars before fermentation can begin. Finally, there are spirits and fortified wines. In the production of these, distillation is used to obtain alcohol concenteration higher than can be achieved by fermentation alone.

**Antibiotics**

Antibiotics production is thought to have a selective value for microorganisms that live in environments where there is intense competition for resources. These are produced by some Ascomycetes and related Deuteromycetes that live in the soil and penetrate and utilise dead plant material. These are also produced by many actinomycetes, especially members of the genus *Streptomyces*. Waksman and Sehatz (1944) to discover an antiboitic Streptomycin from an actinomycete i.e. Streptomyces. Penicillin was first produced for clinical use at Oxford in 1940, from *P. notatum*. In 1945 the Deuteromycete, *Cephalosporium acremonium* was shown to inhibit the growth of both gram positive and gram negative bacteria. A second antibiotic called Cephalosporin N, was found active against both Gram negative and Gram positive bacteria. The majority of antibiotics and produced by microorganisms especially by 20% fungi, 70% actinomycetes and 10% bacteria of isolated antiobiotics belong to these three groups (Berdy, 1974; Glosby, 1976; Lechevalier, 1975).

Griseofulvin from *P. griseofulvum* is one of the relatively few natural organic compound that contains chlorine. It was isolated in 1939 as a result of an investigation into the disappearance of chloride ions from the culture medium. Cyclosporin A, from the deuteromycete *Tolypocladium inflatum,* is a cyclic peptide with many of the constituent amino acids methylated.

**Edible Biomass from Yeasts and Moulds—Single Cell Protein**

Single cell proteins are the dried cells of microorganisms such as as algae, certain bacteria, yeasts, moulds and higher fungi that are grown in large scale culture system for use as protein for human or animal consumption (Davis, 1984; Goldberg, 1985). Much of the world's population is poorly nourished, and famines are frequent. The dietary component that is most usually in short supply is protein. Plants have a relatively low protein content, being largely made of carbohydrates. Plant proteins also tend to be low in some essential amino acids, such as methionine and tryptophan. Animals convert plant biomass into high grade protein, but with poor efficiency, so meat is too expensive for much of the world's population. Many microorganisms are able to use cheap sources of nitrogen such as ammonium salts and nitrates, and abundant carbon sources such as starch, natural gas and petroleum hydrocarbons. The resulting biomass has a high protein content. Microbial growth rates are high and a compact fermentation plant can produce as much protein as a large area of agricultural land. The main objective is supplying protein, the biomass is usually in form of single cell protein or 'SCP' although other cell components will be present, even after processing. Bacteria, fungi and algae have all been used in SCP production but here consideration will be restricted to fungi, both yeasts and moulds. In UK the deuteromycete fungus *Fusarium graminearum* is being used for SCP manufacture from surplus starch. In contrast to most other SCP processes, the product is intended for human consumption. The mycelium stimulates the texture of meat and contributes to the dietary requirement for fibre. It has a high protein content with a satisfactory amino acid content. The lipid content is lower than in meat, with the animal sterol, cholesterol absent. These features have allowed the promotion of this mycoprotein as a health food with the registered trade name 'Quorn'. There is limited production of SCP from cellulose wastes with the help of *Chaetomium* sp. In India, at Central Food Technology Research Institute, Mysore and Indian Agricultural Research Institute, New Delhi, research is being conducted on the use of blue green algae, *Spirulina* as a supplement to diet. The algae is cultured, dried, powdered and then used in the form of capsules or tablets. These contain 60% protein, essential vitamins and unsaturated fatty acids. When fishes and children are fed on this food, encouraging results are obtained. Dabur Research Foundation has recently introduced a health food in the Indian market based on *Spirulina.*

**Biofertilizers**

Biofertilizers are defined as biologically active products or microbial inocultants of bacteria, algae and fungi which may help biological nitrogen fixation for the benefit of the plants. Biofertilizers also include organic fertilizer and manure, which are rendered in an available form due to the interaction of microorganisms or due to their association with plants. Biofertilizers, thus, include the following: (i) symbiotic nitrogen fixers, e.g., Rhizobium spp.; (ii) asymbiotic free nitrogen fixers, such as *Azatobacter-Azospirillium,* (iii) algal biofertilizers, e.g., blue green algae or BGA in association with *Azolla;* (iv) phosphate solubilizing bacteria; (v) mycorrhiza; (vi) organic fertilizers.

Increased usage of chemical fertilizer leads to damage in soil texture and composition, creating environmental problem. Chemicals through the soil enters the food chain and harms the health of human beings and animals. Thus, the use of biofertilizers is both economical and environment-friendly. Government of Inida launched a National Project on Development and Use of Biofertilizers during the Sixth Five Year Plan. Under this project, one national centre and six regional centers and 40 BGA production centres have been established. These centres will produce 800 tonnes of rhizobia and 600 tonnes of BGA annually.

***Rhizobium***

Beijernick in Holland was the first to isolate and cultivate a microorganism from the nodules of legumes in 1888. He named it *Bacillus radicicola* which is now placed in Bergey's Manual of Determinative Bacteriology under the name genus *Rhizobium.* The microbes live freely in soil and in the root region of both leguminous and non-leguminous plants. However, they can enter into symbiosis only with leguminous plants, by infecting their roots and forming nodules on them, the only exception being root nodulation in *Trema* by a *Rhizobium* sp. In legume root nodule symbiosis, the legume is the bigger partner while *Rhizobium* is the smaller partner, often referred to as the microsymbiont. Important species of *Rhizobium* are *R. leguminosarum, R. meliloti* and *R. loti.* Root nodule bacteria have been differentiated, on the basis of growth on a defined substrate, as fast growers and slow growers.

Based on the ability of Rhizobia to produce acid or alkali on yeast extract-mannitol agar medium. *R. phaseoli, R. trifoli, R. leguminosarum* and *R. meliloti* have been grouped as acid producers whereas the slow growing *R. japonicum* and *R. lupin* have been grouped as non-acid producers.

*Rhizobium* can live on relatively simple synthetic media. It has been found that glutamate is much superior to nitrate or the ammonium ion as nitrogen source. It is incapable of fixing atmospheric nitrogen on ordinary media. Rhizobia can fix atmospheric nitrogen and thus not only increase the production of the inoculated crops, but also leave a fair amount of nitrogen in the soil which may benefit the subsequent crop. The infection on legume roots by *Rhizobium* results in the formation of root nodules in which the bacteria may fix $N_2$.

**Asymbiotic Nitrogen Fixers**

Biological $N_2$ fixation is a characteristic property prossessed by a select group of microorganisms by virtue of which molecular

$N_2$ from the atmosphere is converted to a useable form such as ammonia. The special ability of $N_2$ fixing bacteria is reduce $N_2$ to ammonia depends on the possessions of an enzyme system called the nitrogenase complex. It has been shown that 'nif' gene are involved in biological $N_2$ fixation (Christina and Aresa, 1987; Dixon, 1984).

*Azatobacter* and *Azospirillum* when applied to rhizosphere fix atmospheric $N_2$ and make it available to crop plants. They also synthesize growth promoting antibiotic substances helpful to the plant. Most efficient strains of *Azatobacter* fix 30 kg of $N_2$ from 1000 kg of organic matter. When applied to fields, positive responses by field crops were observed leading to saving of 10-25 kg/ha of $N_2$. Similarly, *Azospirillum* with farmyard manure, led to saving of 15-25 kg equivalent of $N_2$ per hectare in crops like sorghum and other millets. *Azospirillum* colonizes not only the roots, but also the above ground portions of the plant through associative symbiosis.

**Algal Fertilizers (BGA + Azolla)**

Blue green algae and *Azolla* constitute a system, which is the main source of algal biofertilizer in south and south east Asia, particularly of low land paddy. BGA inoculation with composite cultures of algal genera, *Anabaena, Nostoc, Plectonema, Aulosira, Oscillatoria* and *Tolypothrix* have been found to be more effective than single cultures. Production and multiplication of BGA cultures is done at various centres in Andhra Pradesh, Uttar Pradesh, Madhya Pradesh, Gujarat, Haryana, H.P. and other states. Application of dried blue green algae flakes at the rate of 10 kg/ha is recommended ten days after transplantation of rice. Besides being a source of $N_2$, BGA provides for the following other advantages: (i) algal biomass accumulates as organic matter; (ii) growth promoting substances become available; (iii) it provides partial tolerance to pesticides and fungicides; (iv) it also helps in reclamation of saline and alkaline soils.

BGA is also supplied with *Azolla* plants which harbour *Anabaena azolle* in leaf cavities, providing symbiotic associates *Azolla* with *Anabaena.* This has been used with some success, but there are following limitations: (i) *Azolla* as a green manuring crop is labour intensive; (ii) raising of *Azolla* needs assured and adequate supply of water; (iii) damage of *Azolla* is caused by pest diseases; (iv) optimum temperature is required for *Azolla* multiplication.

Phosphate solubilising bacteria (e.g., species of *Thiobacillus* and *Bacillus*) and plant growth promoting rhizobacteria (PGPR), including *Pseudomonas fluorescens* and P. *putida,* are important new biofertilizers. PSBs convert non-available inorganic phosphates into soluble organic phosphates which can be utilized by crop plants. PGPRs produce siderophores (iron-chelating substances) and make it unavailable to harmful fungi (*Erwinia*) in rhizosphere, leading to their death.

**Vesicular Arbuscular Mycorrhizae (VAM)**

These mycorrhizas, the commonest, are so called because their haustorial structures within root cells are smooth vesicles or branched arbuscules (tree like structure, from Latin, arbov, a tree) and it is by the presence of these that the mycorrhiza is recognized. The majority, about 90% of all vascular plant species as well as many non-vascular lower plants, have such fungal symbionts, to the extent that mycorrhizas have been described as the main absorbing organs of plants. The group of fungi forming VAMs are placed in order Glomales which is placed in the Zygomycetes on account of their aseptate mycelia and the formation in some species of zygospore-like structures preceded by fusion of two hyphal branches. For convenience, six genera within Endogonales are currently recognized: *Acaulospora, Entrophospora, Gigaspora, Scutellospora, Glomus* and *Sclerocystis,* separated on the basis of spore type and mode of formation. The spores perenate in soil and can be isolated from soil by sieving. The fungus can also survise as mycelial fragments in plant material. Inoculum for agricultural purposes consists of spores, pieces of colonised root or soil.

The six genera named above are responsible for mycorrhizal associations with thousands of different plant species, and single isolate can infect widely different plants. This low host specificity contrasts with that of parasitic biotrophs. A possible explanation for the low specificity is that mutualism, being advantageous to both partners, results in selection for their ability to form associations for and against biotrophs. These VAM fungi can be used commercially (Whetten and Anderson, 1992; Wood, 1992).

Mycorrhizae are mutualistic symbiosis between certain groups of soil fungi and plant root system. In the symbiotic association the fungus utilize carbohydrates produced by the plant. While the plant benefit by the increased uptake of phosphorus and some other nutrients like copper, zinc through the external hyphae extending from the root system into the soil (Bolan, 1991; Koide, 1991). These fungi can increase plant growth under low fertility conditions and are of particular interest of developing low input agriculture system. Numerous reports have shown that mycorrhizal fungi can improve tolerance towards different kinds of stresses such as drought (Giri *et al.*, 1999; Gupta, 1991; Sylvia and Stephen, 1992). These fungi can improve resistance towards root pathogens (Mukerji *et al.*, 1998; Sharma and Mukerji, 1992). Mycorrhizal inoculation will benefit the forest industry (Dixon *et al.*, 1977; Mukerji *et al.*, 1996). There have been many examples of the successful use of mycorrhizal inoculates in improving the establishment or subsequent growth of forest trees or crop plants introduced into new environemnt (Dodd, 1992).

**Orchidaceous Mycorrhizae**

Unlike the VAMs, these are confined to plants of the orchid family, which contains about 15,000 species, mostly tropical. Fungi of orchid mycorrhizas can be grown axenically and are not obligately mycorrhizal as are VAMs. Some are sterile, although many of these show the Basidiomycetous character of clamp connections. Many orchids are associated with *Rhizoctonia* species, including *R. solani* which is also a common plant pathogen. Some of the fungi form fruit bodies enabling them to be identified further. These belong for example to the genera *Corticium* and *Marasmius* and include *Armillaria mellea,* better known as tree pathogen, and *Coriolus versicolor,* a common timber decaying saprotrophy.

**Ericaceous Mycorrhizae**

These, like orchidaceous mycorrhiza, are characterized by association with a specific group of plants, the Ericaceae. This is another large family, with many species and world wide in distribution, which has mycorrhiza of characterstic structure, type of fungus and physiology. Ericaceae form ecologically important plant communities particularly on moss, swamps and on peat, and include plants such as heathers, rhododendrons and azaleas.

**Ectomycorrhizae**

The mycorrhizas are formed on roots of woody plants. A thick fungal sheath develops around the terminal lateral branches of roots and is connected to as intercellular network of hyphae known as the 'Hartig's net' in the root cortex. A wide variety of fungi form ectomycorrhiza but only about 3% plant species support ectomycorrhiza and these are all trees or shrubs. This is presumably because of relatively massive fungal structures involved. The fungal sheath, and in the case of Basidiomycetes, large fruit bodies, can only be sustained by a long term connection with the autotrophic partner. Ectomycorrhizas are believed to be more common in temperatre zones of the world, where there are seasonal climatic changes, than in the tropics. Ectomycorrhizas may affect the ecology of plant communities as well as individual plants. These are thought to have low host specificity.

VAM fungi help in the phosphorus uptake of plants. Phosphorus is one of the most important macro elements for plant life. It occurs as part of DNA and RNA. As part of phospholipid, it constitutes plant membranes. It is part of high energy molecules like ADP. ATP, NADP and NAD which in turn govern all the oxidation-reduction reactions like photosynthesis, respiration, nitrogen metabolism, fat metabolism and other reactions that govern the very existence of plant life.

**Mycoherbicides**

Fungi which help in control of weed growth are known as 'mycoherbicides'. An unwanted plant at a specific place is known as weed. For the control of weeds, it is important to correctly indentify the weed species and to know their biology, mode of dispersal and mode of reproduction. The managment of natural enemies is the basis of biological control of weeds. It is based on the ecological observation that natural enemies can be of prime importance in limiting the distribution and abundance of plants and that some of these natural enemies have restricted host range. Many mycoherbicides are available in the market for commercial use. The mycoherbicide '**Collego**' has been used commercially since 1982 in Arkansas in U.S.A. to control Northern jointvetch (*Aeschynomene virginica*), a leguminous weed in rice (*Oryza sativa*) crop. It is marketed in the form of a dry formulation. '**Devine**' has been used commercially since 1981 in Florida citrus groves to control milkweed vine, *Morrenia oderata.* It is marketed as a wet formulation of chlamydospores of *Phytophthora palmivora* with a shelf life of six weeks in cold storage.

'**Velgo**' is a potential mycoherbicide for velvet leaf (*Abutilion theophrastii*) in corn and soyabean fields in the U.S. corn belt and Southeren Ontaria, Canada. It is based on a strain of *Colletotrichum coccoides.*

'**Luboa-2**' is a selected strain of *Colletotrichum gloeosporioides* f. sp. *cuscuta* that is used as a mycoherbicide to control dodder (*Cuscuta chinesis* and *Cuscuta australis*). '**Biomal**' is a potential mycoherbicide for control of round leaved mallow (*Malva pusilla*). For this weed pathogen is *Colletotrichum gloeosporioides* f. sp. malvae. It is applied in spore suspension containing $2 \times 10^9$ spores at the rate of $5 \times 10^2/1ha^2$. '**Casst**' is another potential mycoherbicide for control of sickel pod, (*Cassia obtusifolia*) by the strain of *Alternaria cassiae* and it applied at the rate of 1.1 kg $ha^{-1}$.

## Degradation of Sewage

The role of microorganism in the decomposition of sewage and other waste material has long been recognized. Conventional sewage treatment involves the use of microorganisms, which develop naturally with in the sewage treatment system. In some newer approaches, however, the sewage is inoculated with a specific microorganism which has been specially selected for that particular sewage treatment process. Such organisms might be called 'Starter cultures'. The use of starter culture increases the efficiency of sewage degradation. A strain of *Pseudomonas putida* containing plasmids has been developed which can degrade octane, xylene, metaxylene and camphor. The degradation of sewage by microorganism requires large amount of oxygen, so that in order to provide room for oxygen most sewage treatment plants require considerable space.

Methane is produced during anaerobic decomposition of sewage and other organic wastes by bacteria. The methane thus produced is collected and used as fuel in many countries. The methanogenic bacteria are able to utilize acetate, methanol, formate, and $H_2 + CO_2$ of the organic wastes for the production of methane gas. Biogas is used run diesel pumps and generators in a meat processing plant. In India cowdung and organic wastes are used. The size of the plant and the quantity of gas produced depend on the quantity of animal waste produced and available. City garbage is also used to produce gas and (Rao, 1986) electricity. Rice husk is also used to produce methane.

**Microbial Control of Disease Vectors**

Microbial control of disease vectors is one of the rapidly growing alternative vector control strategies to control malaria, filaria, Japanese encephalitis, dengue fever and mosquito nuisance. Several species of bacteria, fungi, protozoa and viruses have been evaluated as vector control agents particularly against mosquito larvae. Of these, two bacteria viz., *Bacillus thuringiensis* serotype H-14 and *Bacillus sphaericus*, have been used in vector control on a limitited scale. The two bacteria are known to produce toxic proteins which are specific in action against mosquito larvae. *Bacillus thuringiensis* is toxic to all mosquito species but *B. spharicus* is toxic against *Culex* and some *Anopheles* spp. and non-toxic against *Aedea* spp. Various formulations of two bacteria have been tested. Efficacy of these formulations vary depending upon the type of formulation and various environmental factors. Two bacterial preparations, Bacticite and Sphericide have been evaluated extensively for the control of disease vectors in India.

**Microbes in Metal Cleaning**

Microbiological methods could be effectively used for removal of organic and mineral pollutants from industrial waste waters.

The organic compounds could be utilized by intensifying oxidation reactions by controlling aeration and temperature and also through specific microbial metabolic reactions, such as co-metabolism transformation of functional groups of xenobiotics, etc. The role of bacteria in the leaching of mineral ores was recognised in 1947. When Arthur Colmer and M.E. Hinkle identified a bacterium *Thiobacillus ferroxidans* as the organism principally responsible for the leaching of metal sulphide ore-Bacterial leaching is now being used successfully in many countries throughout the World $L_5$ recover metal from a wide variety of ores (Brierley 1982; Curtin. 1983). Metal ions could be removed by immobilization of products of microorganisms and through oxidation reactions of variable-valency elements. Among the worst group of pollutants of natural environment are heavy metals, mercury, lead, copper, cadmium and zinc. In mining operations, it has been long recognized that *Thiobacillus ferroxidans* and related bacteria produces sulphuric acid from oxidation of reduced sulphur compounds and ferric iron and solubilize metals by a process called bio-leaching. *Thiobacillus* is also used for the recovery of copper and uranium from copper sulphide and uranium oxide respectively.

Bioleaching of oil shales also has the potential to enhance the recovery of hydrocarbons. Many oil shales contain large amounts of carbonates and pyrites and the removal of these minerals increase the porosity of the shale, enhancing recovery of the oil. Acid dissolves the carbonates and can be produced by *Thiobacillus* species growing on the sulphur and Tren in the pyrite.

**Bioremediation**

Microorganisms have dominant roles in such key areas as atmospheric composition, ground water quality, plant and animal productivity, and organic matter cycling. Microbes are also among the most important agents in the biotechnology industry, including genetically engineered microbes that offer hope to degrade hazardous chemicals and to improve plant and animal growth. Bioremediation, a process that exploits the abilities of microbes to enhance the rate or extent of pollution destruction, is an important tool in attempts to mitigate environmental contamination.

**Table 14.2:** Human proteins produced by recombinant micororganisms

| Protein | Product Name | Function and use |
|---|---|---|
| Human growth hormone | Protopin, Humatropin | Hormone that stimulates growth of human body, used in treatment of Drawfism. |
| Insulin | Humulin, Novolin | Hormone that regulates sugar levels in blood used in treatment of Diabetes. |
| Bone growth factor | Somatotropin | Stimulates growth of bone cell, used in treatment of osteoprorosis. |
| Interferon alpha | Berofor, Intron A., Wellferon, Referon A., human recombinant alpha interferon | Cytokine of immune system used in treatment of cancer and viral diseases. |
| Interferon Beta | Frone, Betaseron, human recombinant beta interferon | Cytokine of immune system used in treatment of cancer and viral diseases. |
| Interferon gamma | Actimmune | Lymphokine of immune system, used in treatment o cancer and viral diseases. |
| Interleukin 2 | Proleukin, human recombinant interleukin-2 | Lymphokine of immune system that stimulates T cell used in Treatment of immune deficiencies and cancer disease. |

(*Contd.*)

| Protein | Product Name | Function and use |
|---|---|---|
| Tumor necrosis factor (TNF) | — | Lymphokine of immune system that causes death of malignant cells used in treatment of cancer. |
| Tissue plasminogen activator (TPA) | Actilyse | Dissolve blood clots, used in treatment of heart disease and during heart surgery. |
| Blood clotting factor VIII | Recombinate | Stimulates blood clot formation, used in treatment of hemophiliaes. |
| Epidermal growth factor | — | Regulates calcium levels and stimulates growth of epidermal cells, used in treatment of wounds to stimulate healing. |
| Granulocyte colony stimulating factor | Filgrastin Neupogen | Regulates production of neutrophils in bone marrow, used in treatment of cancer to prevent infections. |
| Erythropoietin (EPO) | Procrit, Epogen | Stimulates red blood cell production, used in treatment of anemia in dialysis patient. |

Bioremediation achieves contaminant decomposition or immobilisation by exploiting the existing metabolic potential in microbes collectively.

## Microbes in Medicine

Biotechnology also helps in producing efficaceous drugs are also manufactured for treatment of diseases. Insulin and interferons are two such drugs which help in controlling the diabetes and some tumour viruses respectively. Such drugs can be manufactured now in bacterial cell cultures in large quantities, if the corresponding genes from human being or animals are cloned through plasmid vectors in bacteria. This making their production relatively very cheap. The gene for insulin was cloned in bacteria and could be used for synthesis of insulin. Dr. Saran Narang, a scientist of Indian origin, working in Ottawa, Canada was involved in cloning of insulin gene. Synthetic insulin manufactured in this manner is now being sold commercially in North America. About a dozen different interferons are also in different stages of testing and some of them are already being sold commercially. In 1985, human growth hormone for treating hypopitiutary dwarfism was synthesized using recombinant DNA technique and was approved for commercial marketing under the name 'Prototropin' USA and under the name 'Somatonorn' in Britain (Tabel 14.2).

## Transgenic Crops

Transgenic canola, maize, cotton, potato, soybean, squash and tomato varieties are available commercially in North America. A large part of the transgenic crops are expressing genes from *Bacillus thuringiensis* (*Bt*) for the production of crystal proteins in plant tissues to resist insects. The introduction of Bt crops have a positive impact, most importantly the potential reduction in chemical insecticide use. Cotton, for example has received traditionally high levels of insecticide applications of which the high economic, environmental and human health costs have been well documented. Infact, the availability of Bt cotton may permit the reintroduction of the crop into those parts of the region that have stopped producing cotton due to the high costs of insecticides. A second potential benefit of transgenic Bt crops the easy of their implementation.

*Bacillus thuringiensis* (*Bt*) toxin genes have been genetically engineered into dozens of plant species. By expressing a Bt toxin with in their tissues, the plants protect themselves from some insect pests without farmers resorting to pesticide sprays.

Most living cells are protein factories that manufacture products according to the genetic code with in them. Through the use of gene splicing technique bacterial cells can be altered so that they will turn out a product encoded by a foreign gene that has been spliced into a bacterial plasmid. Such altered bacteria are then grown in large scale fermentation tanks or bioreactors.

In India during the last decades or so with the efforts of Government of India many Centralized service oriented research facilities have been established many new programmes have been introduced to generate highly skilled and trained workers. This has strengthened the base of biotechnology and given a big momentum to Research and Development activities in these existing field.

# LITERATURE CITED


Ainsworth, G.C. 1971. Ainsworth's and Bisby's Dictionary of the Fungi. 6 Ed. Common Wealth. Mycological Institute, Kew Surrey.

Alexopoulous C.J., Mims, C.W. and Blackwell, M. 1996. Introductory Mycology, John Wiley & Sons. pp. 1-868.

Alexopoulous, C.J. 1960. Gross morphology of the plasmodium and its possible significance in the relationships among the myxomycetes. Mycologia 52 : 1-20.

Aronson, J.M. 1965. The cell wall, In : The Fungi : Vol. I (Eds. G.C. Ainsworth and A.S. Sussuman) Academic Press, New York.

Atkinson, R.G. and Robinson, J.B. (1955). The application of a nutritional grouping method to soil fungi. *Canadian Journal of Botany* 33, 281-288.

Ballentine, R. and Stephens D.G., 1951. J. Cellular Comp. Physiol. 37, 369-387. .

Barendanm, W. 1928. *Abstract II*, 75, 426-452, 503-533.

Barghoorn, E.S. and Linder, D.H., 1944. *Forlowia.* 1, 395-467.

Bennett, C.W. 1921. *Michigun State University Agricultural Experimental Station Technological Bulletin*. 53, 1-40.

Benjamin, R.K. 1979. In : The whole Fungus (Ed. W.B. Kendrick) Vol. II. National Museum of Canada, Ottawa, pp. 573-621.

Benjamin, R.K. 1979. Zygomycetes and their spores. Ch. 23. In W.B. Kendrick (ed.) the whole Fungus National Museums of Canada, Ottawa.

Benjamins, R.K. 1979. Zygomycetes and their spores. In : The whole Fungus, Vol. 2. (Eds B. Kendrick). National Museums of Natural Science, Ottawa, pp. 573-621.

Berch, S.M. 1986. Endogonaceae : Taxonomy, specificity fossil record, phylogeny ; In Frontiers in Applied Microbiology; Vol. 2, pp. 161-186 eds. (K.G. Mukerji, N.C. Pathak and V.P. Singh) Lucknow, India, India Print House.

Bhandari, N.N. and Mukherji, K.G. 1993. The Haustorium, John Wiley and Sons Ltd. U.K.

Behr, G. (1930). *American Journal* of Botany. 1, pp. 418-444.

Berdy, J. 1974. Recent developments of antibiotic research and classification of antibiotics according to chemical structure *Adv. Appl. Microbiol.* 18: 309-406.

Boic, D. 1925. Uber die Chemischen Character der Peridie, des kapillitiums und der sporen membraner bei Myxomyceten. Aeta. Bot. Inst. Bot. Univ. Zagreb. 1 : 44-63.

Brierley, C.L. 1982. Microbial mining, *Scientif. Amer.* 247: 42.

Brierley, W.B., Jewson, S.T. and Brierley, M. 1928, *Proccedings of International Congress on Soil Science.* 3, 48-71.

Brooks, S.C. and Brooks M.M., 1941. The Permeability of Living cells. *Berlin Zehlendorf. Gebruder Borntraeger*. pp. 395.

Bruns, T.D. White T.J. and Taylor, J.N. 1991. Fungal Molecular Systematics. Ann. Rev. Ecol. Syst. 22 : 525-564.

Buell, C.B. and Weston W.H. (1947). *American Jounral of Botany,* 34 : 55-561.

Burger, A. 1939. Broteria, Ser. trimestral. 8 : 64-81.

Cale, G.T., and Kendrick, W.B. 1968. Conidium Ontogeny in Hyphomycetes. The imperfect state of *Monascus* ruber and its meristem arthrospores. Can J. Bot. 46 : 987-992.

Cantino, E.C. (1949). *American Journal of Botany*. 36, 95-112.

Christina, K. and Aresa, J. 1987. Genetics of Azatobactors: Cation of $N_2$ fixation and related aspects of metabolism. *Ann. Rev. Microbiol.* 41: 227-258.

Cochrana, V.W. 1958. The Physiology of Fungi XIII + 524 pp. John Wiley, New York.

Crasemann, J.M. 1954. *American Journal of Botany* 41, 302-310.

Crueger, W. and Crueger, A. 1982. *Biotechnology: A Textbook of Industrial Microbiology*, ed. T.D. Book. *Sci. Tech.* Inc. Madison, Wisconsin, USA.

Curtin, M.E. 1983. Microbial mining and metal recovery: Corporation take the long and cautions path. *Biotech.* 1: 229.

Davis, P. 1984. *Single Cell Protein.* Academic Press, New York, USA.

Day, D. and Harvey A., 1946. *Plant Physiology.* 21, 233-236.

De. Bary, A. 1887. Comparative morphology and biology of the fungi, mycetozoa, and bacteria. Trans by H.E.F. Garnsey Revised by L.B. Balfour XIX + 525 pp. 198 figs. Clarendon Press, Oxford.

DeBary, A. 1859. Die Mycetozoen. Ein beitrag zur kenntnis der Niedersten Thiere. Zeits. Zool. 10 : 88-175.

Denny, F.E. (1933). Contribs. *Boyce Thompson Institute.* 5 : 95-102.

Dimmling, W. and Neseman, G. 1985. Critical assessment of feed stocks for biotechynology, *CRC Crit. Rev. Biotech.* 2: 233: 285.

Dixon, R.A. 1984. The genetic complexity of $N_2$ fixation. *J. Gen. Microbiol.* 130: 2745-2755.

Dodel, J.C. 1992. The use of mycorrhizal technology in sustainable agro ecosystem. *J. Sci. Food. Agric.* 60: 396.

Domsch, K.H. (1960). Das Pilzspektrum other Bodenprobe. II. Nachweis physiologischer merkmale. *Archiv for Mikrobiologie.* 35, 229-247.

Elliot, E.W. 1949. The swarm cells of Myxomycetes, Mycologia 41 : 141-170.

Emerrson, W.W. (1959). The structure of soil crumbs. *Journal of Soil Science* 10, 235-244.

Foster, J.W. 1949. Chemical activities of Fungi XVII! + 648 pp. Academic Press, New York.

Garrett, S.D. (1951). Ecological groups of soil fungi; a survey of substrate relationships. *New Phytologist*, 50, 149-166.

Gaumann, E. and U Nef. 1947. Ber - Schweiz botan. Ges. 57, 258-271.

Gerdemann, J.W. and Nicolson, T.H. 1961. Spores of mycorrhizal *Endogone* species extracted from soil by wet sieving and decanting. Trans. Brit. Myol. Soc. 46 : 235-244.

Gerdemann, J.W. and Trappe, J.M. 1974. The Endogonaceae in the Pacific North West, Mycologia Memoir No. 5, pp.65.

Gerdemann, J.W. and Trappe. J.M. 1975. Taxonomy of the Endogonaceae. pp. 35-51. In. F.E. Sanders. B. Mosse. and P.B. Tinker (eds) Endomycorrhizas. Academic Press, New York.

Golding, N.S. 1940. *Journal of Dairy Science*, 23, 879-889.

Greene, H.C. and Fred. E.B., (1934) *American Journal.* 26 : 1297-1299.

Griffths, D. 1901. The North American Sordariaceae. Torrey. Bot. Club Man II 134 pp.

Gupta, R. and Mukerji, K.G. 2001. Microbial Technology APH. Publishing Company pp. 1-206.

Gupta, V.C. (1967). Carbohydrates. In Soil Biochemistry, A.D. McLasen and G.F. Peterson, eds., : Marcel Dekker Inc., New York, pp. 99-118.

Hall, I.R. 1983. A summary of the Features of Endogonaceous Taxa; Tech. Rep. No. 8. Agricultural Research Centre, Mosigel, New Zealand.

Harper, J.E. and Webster, J. (1964). An experimental analysis of the Coprophilous fungus association. *Transactions of the British Mycological Society.* 47, 511-530.

Hayashi, K. 1954. *Japanese Journal of Botany*. 14, 91-98.

Hildebrand, E.M. (1938). *Botanical Review*, 4, 627-664.

Hosten, B. Von, A., Von, Hofsten and Fries N., 1953, *Experimental Cell Research*. 5, 530-535.

Horowitz, N.H. 1951. Growth 15 Supplement. : 47-62.

Jermyn, M.A. 1953. *Australian Journal of Biological Science*, 6 : 48-69.

Jilson, O.F.and Nickerson W.J., 1948. *Mycologia.* 40, 369-385.

Kaeses, G. and Schwartz W., 1935 *Arch. Mikrobio*l. 6, 208-214.

Kendrick, W.B. and Carmichael, J.W. 1973. Hyphomycetes. pp. 323-509. In G.C. Ainsworth, F.K. Sparrow, and A.S. Susswan (eds). The Fungi Vol. IV A Academic Press, New York.Lewis, J.A. and Starkey, R.L. (1969). Decomposition of plant tannins by some microorganisms. Soil Science. 107, 235-241.

Koehn, R.D. 1971. Laboratory culture and ascocarp development of *Podosordaria leporina*. Mycologia 63 : 441-458.

Lecheratier, H. 1975. Production of some antibiotics by members of different genera of microorganisms. *Adv. Appl. Microbiol.* 19: 25-45.

Lichtwardt, R.W. 1986. The Trichomycetes. Springer Verlag, New York.

Lundberg, J.G. and McDade, L.A. 1990. Systematics In : Methods for Fish Biology Eds (C.B. Schreck, and P.B. Moyle) American Fisheries Society, Bethesda, MD. pp. 65-108.

Manoharachary, C. Kunwar, I.K. and Mukerji, K.G. 2002. Arbuscular Mycorrhizal Fungi—Identification, Taxonomic Criteria, Classification, Controversies and Terminology. In : Mukerji et. al. (eds.) Techniques in Mycorrhezial Studies. Kluwer Academic Publishers, Dordrecht, pp. 249-272.

Martin, G.W. 1960. The systematic position of the myxomycetes. Mycologia 52 : 119-129.

Mayr, E. 1942. Systematics and the Origin of the species Columbia University Press, New York.

Mayr, E. 1983. Animal species and Evaluation. Harvard University Press, Cambridge, USA.

Mayer, W.V. 1986. The dawning age of biotechnology. In: *Biological Technology and Biological Education.* Yarmonk University, Amman, Jordan.

Mehta, K.C. 1929, "Annual occurrence of rusts on wheat in India." Presidential address (Sec. of Botany) Proc. 16th Indian Sci. Congr. pp. 199-223.

Mehta, K.C. 1940, "Further studies on cereal rust in India," Sci. Monograph Imperial Com. Agr. Res. India, 14 : 1-244.

Mehta, K.C. 1952. Further studies on cereal rusts in India II Indian com. Agr. Res. Sa. Mon. 18 : p. 368.

Moore, R.T. and Mc Clear, J.H. 1962. Fine structure of mycota 7. Observations on septa of Ascomycetes and Basidiomycetes. Am. J. Bot. 49 : 86-94.

Morton, J.B. 1990. Evolutionary relationship among arbuscular mycorrhizal fungi in the Endogonaceae, Mycologia, 82 : 192-207.

Morton, J.B. and Benny, G.L. 1990. Revised classification of arbuscular mycorrhizal fungi (zygomycetes) : a new order, Glomales, two new sub orders Glomineae and Gigasporineae, and two new families. Acualosporaceae and Gigasporaceae with an emendation of Glomaceae, Mycotaxon 37 : 471-491.

Moss, S.T. and Young, T.W.K. 1978. Phyletic Considerations of the Harpellales and Asellariales. Mycologia 70 : 944-963.

Moss, S.T., Lichtwardt, R.W. and Manier, J.F. 1975. Zygopolaris a New Genus of Trichomycetes Producing Zygospores with Polar attachment. Mycologia 67 : 120-127.

Moyer, A.J. and Coghill R.D., 1946, Journal od Bacteriology. 51, 57-78.

Meyer, E. (1955) *Mycologia.* 47, 664-667.

Mukerji, K.G., Chamola, B.P., Kaushik, A., Sarkar, M. and Dixon, R.K. 1996. Vesicular arbuscular mycorrhiza: Potential biofertilizer for nursery raised multipurpose tree species in tropical soils. *Ann. For.* 4: 12-20.

Mukerji, K.G., Chamola, B.P. and Sharma, M. 1998. Mycorrhiza in control of plant pathogens. In : *Management of Treatening Plant Diseases of National Importance,* eds. V.P. Agnihotri, A.K. Sarbhoy and D.V. Singh. Malhotra Publishing House, New Delhi, pp. 297-314.

Mukerji, K.G. 1996. Taxonomy of endomycorrhizal fungi in : Advances in Botany, K.G. Mukerji, B. Mathur, B.P. Chamola and P. Chitralakha, eds. A.P.H. Publishing Corporation, New Delhi, pp. 212-219.

Mukerji, K.G. and Dixon, R.K. 1992. Mycorrhizae in reforestation. In : Proceedings International Symposium on Rehabilitation of Tropical Rain forest Ecosystems, Research and Development. (Ed. A.B. Said) Sarawak, Malaysia pp. 66-82.

Mukerji, K.G. and Kapoor, A. 1986. Occurrence and importance of vesicular arbuscular mycorrhizal fungi in semiarid regions of India. For Ecol. Manag. 16 : 117-126.

Mukerji, K.G., Manoharachary, C. and Chamola, B.P. (eds.) 2002. Techniques in Mycorrhizae Studies. Kluwer Academic Publishers, Dardrech, p. 554.

Nabel, K. 1939. Uber die Membran niederer Pilze, besonders Von Rhizidomyces bivellatus nov. spez. Asch. Mikrobiol. 10 : 515-541.

Nickerson, W.J. and Edwards G.A., 1949. *Journal of Ge-Physiol.* 33, 41-55.

Nicholas, D. J.D. (1965) Utilization of inorganic nitrogen compounds and amino acids by fungi. In: The Fungi, Vol. 2. G.C. Ainsworth and A.S. Sussman eds., Academic Press, New York, pp. 349-376.

Norstadt, F.A. and McCalla, T.M. (1969) Microbial populations in stubble mulched soil. *Soil Science*. 107, 188-193.

Old, K.M. (1969). Perforation of conidia of *Cochlibolus sativus* in natural soils. *Transactions of the British Mycological Society,* 53, 207-216.

Paper, K.B. and Fennell, D.I., 1965. *The Genus Aspergillus.* The William and Wilkins Co., Baltimore, USA. 686 p.

Pirozynski, K.A., and Dalpe, Y. 1989. Geological History of the Glomaceae with Particular Reference to Mycorrhizal Symbiosis 7 : 1-36.

Pontecorvo, G. 1956. The parasexual cycle in fungi Ann. Rev. Microbiol. 10 : 393-400.

Rao, A.M., 1982. Certain aspects of biotechnology with emphasis on plant tissue culture. In : *Biological Technology and Biological Education.* Proc. of the IUBS, Yormonk University, Amman, Jordan.

Raper, J.R. 1966. Genetics of Sexuality in higher fungi. Ronald Press, New York. 283. pp.

Remy, W., Taylor, T.N., Hass, H. and Kerp, H. 1994. 400 million year old vesicular arbuscular mycorrhizae (VAM), Proc. Nat. Acad. Sci. USA 91 : 11841-11843.

Schenck, N.C. and Perez, Y. 1990. Manual for the identification of VA mycorrhizal fungi VAM, 3rd edn; (Florida, USA : University of Florida).

Schwendener, S. 1867. Abh. Schweiz, Natur horsch Ges, 51 : 86-90.

Sharma, M. and Mukerji, K.G. 1992. Mycorrhiza—Tool for biocontrol. In : *Recent Development in Biocontrol of Plant Diseases*, eds. K.G. Mukerji, J.P. Tewari, D.K. Arora and G. Saxena, Aditya Books Pvt. Ltd. New Delhi, pp. 52-80.

Sherf, A.F. (1943) *Phytopathology*, 33, 330-332.

Smith, E.C. 1929. Longevity of myxomycete spores. Mycologia 21 : 321-323.

Sparrow, F.K. 1960. Aquatic Phycomycetes XXV + 1187 pp. Univ. of Michigan Press. Ann Arbor.

Stevens, R.B. 1974. Mycology Guide book. 703 pp. Univ. of Washington Press, Seatle and London.

Stockdale, P.M. (1953) *Journal of Gen Microbiology*. 8, 434-441.

Sylvia, D.M. and Stephan, E.W. 1992. Vesicular arbuscular mycorrhiza and environment stress. In: Mycorrhizae in Sustainable Agriculture, eds., G.J. Bethelenfalvay and R.G. Linderman, American Society of Agronomy, special Publication No. 54 pp. 101-124.

Talbot, P.H. B. 1971. Principles of Fungal Taxonomy. St. Martins Press, New York, pp. 274.

Tamiya, H. 1929. *Acta Phytochem* (Japan). 4, 227-295.

Towe, K.M. 1981. Biochemical keys to the emergence of complex life. In : Life in the Universe (Ed. J. Billingham). MIT Press, Cambridge, MA.

Trappe, J.M. 1987, Phylogenetic and ecological aspects of mycorrhiza in the angiosperms from an evolution stand, piont in : Ecophysiology of VA mycorrhizal plants, G. Safir, ed. CRC Press, Boco Raton, Florida, pp. 5-25.

Tribe H.T. (1960). Aspects of decomposition of cellulose in canadian soils. I. Obsrevations with the microscope. *Canadian Journal of Microbiology*, 6, 309-316.

Vishniac, H.S. 1955. *Mycologia*. 47, 633-645.

Walker, C. and Trappe, J.M. 1992. Names and epthets in Glomales and Endogonales. Mycol. Res. 97 : 339-344.

Walker, C.M. 1992. Systematics and taxonomy of the arbuscular endomycorrhizal fungi (Glomales) a possible wing forword, Agronomic Agronomic, 12: 887-897.

Whetten, R. and Anderson, A.J. 1992. Theoretical consideration in the commercial utilization of mycorrhizaal fungi. In : Handbook of Applied Mycology, Vol. IV, *Fungal Biotechnology*, eds. D.K. Arora R.P. Elander and K.G. Mukerji. Marcel Dekker Inc. New York, USA, pp. 849-879.

Whittaker, R.H. 1969. New concepts of kingdoms of organisms. Science 163 : 150-160.

Wilem, S. (1955) *Phytopathology*, 45. 180-181.

Wood, T. 1992. Vamycorrhizal fungi : Challenges for Commercialization. In : *Handbook of Applied Mycology*, Vol. IV, *Fungal Biotechnology*, eds. D.K. Arora, R.P. Elander and K.G. Mukerji. Marcel Dekker Inc., New York, USA, pp. 823-848.

Zoble, K.H. (1943). Arch. Microbiol. 13 : 191-206.

# Subject Index

A

**D**

**H**



**I**

K



L



M

N



O



P

**T**



**U**



**V**

**W**



**X**



**Y**



**Z**